Unbounded Non-Commutative Integration

MATHEMATICAL PHYSICS STUDIES

A SUPPLEMENTARY SERIES TO
LETTERS IN MATHEMATICAL PHYSICS

Editors:

J. C. CORTET, *Université de Dijon, France*
M. FLATO, *Université de Dijon, France*
M. GUENIN, *Institut de Physique Théorique, Geneva, Switzerland*
E. H. LIEB, *Princeton University, U.S.A.*
R. RACZKA, *Institute of Nuclear Research, Warsaw, Poland*

Editorial Board:

W. AMREIN, *Institut de Physique Théorique, Geneva, Switzerland*
H. ARAKI, *Kyoto University, Japan*
A. CONNES, *I.H.E.S., France*
L. FADDEEV, *Steklov Institute of Mathematics, Leningrad, U.S.S.R.*
J. FROHLICH, *F.T.H., Switzerland*
C. FRONSDAL, *UCLA, Los Angeles, U.S.A.*
I. M. GELFAND, *Moscow State University, U.S.S.R.*
A. JAFFE, *Harvard University, U.S.A.*
A. A. KIRILLOV, *Moscow State University, U.S.S.R.*
A. LICHNEROWICZ, *Collège de France, France*
B. NAGEL, *K.T.H., Stockholm, Sweden*
J. NIEDERLE, *Institute of Physics CSAV, Prague, Czechoslovakia*
A. SALAM, *International Center for Theoretical Physics, Trieste, Italy*
W. SCHMID, *Harvard University, U.S.A.*
I. E. SEGAL, *M.I.T., U.S.A.*
J. SIMON, *Universityé de Dijon, France*
D. STERNHEIMER, *Collège de France, France*
I. T. TODOROV, *Institute of Nuclear Research, Sofia, Bulgaria*

VOLUME 7

Unbounded Non-Commutative Integration

by

J. P. Jurzak

Physique-Mathématique, Université de Dijon, France

D. Reidel Publishing Company

A MEMBER OF THE KLUWER ACADEMIC PUBLISHERS GROUP

Dordrecht / Boston / Lancaster

Library of Congress Cataloging in Publication Data

CIP

Jurzak, J. P. (Jean-Paul), 1950–
Unbounded non-commutative integration.

(Mathematical physics studies ; v. 7)
Bibliography: p.
Includes index.
1. Integration, Functional. 2. Von Neumann algebras.
3. Noncommutative algebras. 4. Operator theory. 5. Mathematical physics. I. Title. II. Series.
QC20.7.F85J87 1985 530.1′5157 85-10759
ISBN 90-277-1815-6

Published by D. Reidel Publishing Company
P.O. Box 17, 3300 AA Dordrecht, Holland

Sold and distributed in the U.S.A. and Canada
by Kluwer Academic Publishers,
190 Old Derby Street, Hingham, MA 02043, U.S.A.

In all other countries, sold and distributed
by Kluwer Academic Publishers Group,
P.O. Box 322, 3300 AH Dordrecht, Holland

Printed in The Netherlands

À DOMINIQUE

TABLE OF CONTENTS

PREFACE

Non-commutative integration has its origin in the classical papers of Murray and von Neumann on rings of operators, and was introduced because of unsolved problems in unitary group representations and the elucidation of various aspects of quantum-mechanical formalism, together with formal calculus in such operator rings. These papers emphasized the interest in II_1-factors and pointed out the remarkable behavior and algebraic structure of the set of all unbounded closed operators affiliated to such rings. The absence of power tools in functional analysis - mainly settled in their definitive form by A. Grothendieck around 1950-1955 - together with the pathological manipulation of algebraic operations on closed operators in Hilbert spaces, has limited ring-theory to the study of algebras of bounded operators with the main objective the difficult question of classification up to isomorphisms of factors.

This material has permitted a rigorous study of discrete systems in statistical mechanics but appears to be less convincing in other domains of physics (in the algebraic approach to field theory, for example). The striking role of Hamiltonians, Schrödinger operators and Lie group invariant properties in such areas of physics disappears in the so-called $\mathbb{C}^*$-approach. One has to note here that algebras of bounded operators may efface certain features: in this respect we find, for a space A of observable containing all powers Δ^n of a given essentially self-adjoint hamiltonian Δ, assuming that Δ^{-1} is compact, that states on A are associated to particular trace class operators acting in the Hilbert space H - even if A contains the space $L(H)$ of all bounded operators in H - i.e., the predual of A is identical to its strong dual. Self-adjointness of hamiltonians is a natural physical property, though we know that general symmetric observables may have zero self-adjoint extension. Finally, the presence of quadratic forms seems unavoidable in these theories and has been already introduced by certain authors.

However, many mathematical problems in Hilbert spaces cannot be reduced to bounded operators problems, even with the use of bounded functions (when defined?) of unbounded operators. It is significant that Sobolev spaces, or related spaces of partial differential equations, are constructed from self-adjointness on C_c^∞ of a suitable dominating partial differential operator (one may take all powers of the

Laplacian) and these spaces of regular vectors lead to the manipulation of the supple space of differentiable vectors for a self-adjoint operator or, more generally, of the space of C_c^∞-vectors for a Lie group representation canonically present. A common denominator in these situations is the precense of a Fréchet space topology on these domains of regular vectors, which dictates the behavior of these dominating operators and explains the methods of estimations a priori. Such Fréchet spaces are often Schwartz spaces or nuclear spaces of C^∞-functions (a consequence of the Rellich theorem and Sobolev lemma) and correspond to the discrete nature of the spectra of the operators mentioned. These techniques also work in enveloping algebras obtained from the differentiation of a Lie group representation, where domination is ensured by the image of the Laplacian constructed on this group G. Finally, the Plancherel formula on G is a situation involving the distribution at the neutral element and cannot be treated with bounded operators only.

The fusion of all the preceding ideas and situations has led the author to modify the ring-operator theory in order to make it applicable to these rough contexts. This monograph is devoted to the study of the indispensable tools needed to accomplish this. The analogue of the Hamiltonian or of the Laplacian must be a fixed self-adjoint operator Δ whose property is that all its powers Δ^n are dominating in the space considered. The material so obtained is different, and not equivalent, to von Neumann's theory, though essentially the result of von Neumann investigations.

The author thanks Professors H. Araki and M. Flato for their kind interest in this work and valuable comments.

INTRODUCTION

This work develops the material for unbounded non-commutative integration and describes the relation between these developments and material in analysis concerned with regularity questions. Unbounded operators automatically appears in such questions and their manipulation is known to be delicate in limit processes, commutation questions, etc.

On the other hand, traditional non-commutative integration is a well-established theory dealing with algebras of bounded operators and has been distant from classical analysis for a long time; recently, however, important applications have been developed by A. Connes (see [5]).

The commutative case is clearly concerned with functional integration and makes the use of unbounded measurable functions transparent. In an unbounded non-commutative case, they are replaced by continuous sesquilinear forms on a given dense domain $\mathcal{D}$ - endowed with a natural Fréchet topology - dense in some Hilbert space H; note that a closeable operator T in H with a domain containing $\mathcal{D}$ is viewed as the continuous sesquilinear form $(x, y) \in \mathcal{D} \times \mathcal{D} \to (Tx, y)$. Elements of the dual or antidual of $\mathcal{D}$ are sometimes called distributions, as it is clearly seen in example 6. A known example is $\mathcal{D}$ equal to the Fréchet space of C^∞-vectors for a Lie group representation. The usual Banach algebra $L(H)$ is replaced - except in commutation questions - by the set $B(\mathcal{D}, \mathcal{D})$ of all continuous sesquilinear forms β on the product space $\mathcal{D} \times \mathcal{D}$, and, as $B(\mathcal{D}, \mathcal{D})$ is the dual of the (Fréchet space) projective tensor product $\mathcal{D} \hat{\otimes} \mathcal{D}$ (isomorphic to the predual of $B(\mathcal{D}, \mathcal{D})$), we see how smooth the limit process is for sequences or nets in $B(\mathcal{D}, \mathcal{D})$.

The concept of a von Neumann algebra is replaced by the concept of an ultraweakly closed space A with condition II; in general, A is not an algebra and one may consult the paragraph on examples and comments below. The bounded part A_{id} (i.e. the set of β in $\mathcal{D} \times \mathcal{D}$ which extends continuously to $H \times H$) is a von Neumann algebra contained in A practised from many points of view.

The strong dual A^ρ and the predual space P_A of A are always Fréchet spaces and the space A is endowed with an inductive limit topology - called ρ - of normed spaces A_{A_i} constructed from domination with a given $A_i \geq Id$ in A; this domination is closed to Nelson's domination in [15] and to generalized strong convergence studied in [12]. The passage from A_{A_i} to A_{id} is ensured by an easy lift process corresponding to a homeomorphism of A. This paper is a continuation of [1, 10] and has been written mainly in a self-contained manner.

The aim of Chapter 1 is to make the reader familiar with ultraweakly closed spaces (with condition II) and commutation theory inside $B(\mathcal{D}, \mathcal{D})$, which is sufficient for many questions. Our point of view is slightly different from [1] - though proof of theorem 1.2 is directly adapted from this paper - and try to present an ultraweakly closed space A as a natural object existing alone independently of any $*$-algebra in it. Note also that excessive generality may be avoided by considering those spaces A with a cofinal abelian sequence (see also examples 2, 4 and remark 1.1.1°/). We also see that A may be viewed as the space of elements β in $B(\mathcal{D}, \mathcal{D})$ affiliated to the von Neumann algebra $M = A_{id}$.

Chapter 2 essentially establishes the density of M for ρ-topology considered on A.

Chapter 3 concerns the technical results on elementary operations (reduction, induction, etc.) as a preparation for Chapter 10.

Chapter 4 is devoted to a detailed analysis of the Fréchet space $\mathcal{D}$ its strong dual $\mathcal{D}'$ with a supplementary section on the manipulation of closed operators in Hilbert spaces. The construction of the domains necessary for a general commutant theory is indicated (compare with example 3).

Chapter 5 describes the abelian case. Using remark 1.1.1°/, we are reduced to $A = \cup_{n\geq 0} A_{\Delta^n}$, $\Delta \geq 1$: the image of the Gelfand transformation is explicit and the set X of characters of A is a dense open set of the compact space K (K is the space of characters of A_{id}) thus representing elements of A as continuous functions on X. Ultraweakly closed abelian A are precisely described in example 1 and we extend the usual duality $\langle L^1, L^\infty \rangle$ to this situation. Theorem 5.3 is of special importance.

Chapter 6 is an illustration of the technicalities and may be omitted upon first reading. Spaces A, in direct connexion with von Neumann algebra theory, are those for which A_i^{-1} is central (for a von Neumann algebra, one has $A_i = Id$) and are easily handled. In particular, all classical results of $\mathbb{C}^*$-theory concerning derivations are put in this context.

Chapter 7 is concerned with the notion of a representation. A representation π of A is a linear map from A into the space $B(\mathcal{D}_\pi, \mathcal{D}_\pi)$ of continuous sesquilinear forms on another domain $\mathcal{D}_\pi$ dense in some Hilbert space H_π: the restriction of π to the involutive algebra A_{id} - i.e., the bounded part of A - is a representation of this algebra in the Hilbert space H_π and π acts on the $*$-algebra $\mathcal{P}$ generated by all A_i, A_i^{-1} as a representation from $\mathcal{P}$ into an involutive algebra of operators sending $\mathcal{D}_\pi$ into $\mathcal{D}_\pi$; one has $\mathcal{D}_\pi = \cap_{j\geq 0}\, \pi(A_j^{-1}) H_\pi$, $\pi(A_i^{-1})^{-1} = \overline{\pi(A_i)}$ and $\pi(A_i)^{\varepsilon}\, \mathcal{D}_\pi = \mathcal{D}_\pi$ for $\varepsilon = \pm 1$ and all $i \geq 0$; the image of $\beta = BA_i \times A_i \in A$ where $B \in A_{id}$ is $\pi(\beta) = \pi(B)\pi(A_i) \times \pi(A_i) \in B(\mathcal{D}_\pi, \mathcal{D}_\pi)$. This definition appears both supple and stable under known operations such as subrepresentations, direct sums, tensorproducts, etc.

The case $\mathcal{P}$ abelian is essential (note the analogy with maximal abelian subalgebra of semi-simple complex Lie algebras and with Cartan subalgebra iff characters on $\mathcal{P}$ are normal); in particular, it is shown that a positive linear form f on A induces a G.N.S. representation π_f of A with the following properties:

1°/ the restriction π_f to A_{id} is the usual G.N.S. representation - with cyclic vector ζ_f in Hilbert space H_f - associated to the couple (A_{id}, f);

2°/ One has $\zeta_f \in \cap_{j\geq 0}\, \pi(A_j^{-1})(H_f) \equiv \mathcal{D}_f$ and $f(\beta) = (\pi(\beta)\zeta_f, \zeta_f)$ for all $\beta \in A$.

For applications, it seems sufficient to deal with A ultraweakly closed and $f \geq 0$ normal on A. Note that the map

$$T \in A \cap L(\mathcal{D}) \longmapsto j(T) = \pi(T)\pi_f \in \mathcal{D}_f$$

extends continuously from A into $\mathcal{D}_f'$, i.e., for $\beta \in A$, $j(\beta)$ is a distribution on $\mathcal{D}_f$.

We include, as a supplement to Chapter 7, a paragraph on the second dual of A objects [10]; the results are natural generalizations of W*-theory. In particular, the second dual of a space A, with condition II and cofinal abelian sequence A_i, is an ultraweakly closed space with condition II and a cofinal abelian sequence.

Chapter 8 states all properties in direct connexion with the strong dual A^ρ of A. For a space A with condition II, we introduce the convex set S of linear forms $f \geq 0$ on A, such that $f(1) = 1$. This set S imbeds into the compact set K - with $\sigma(A_{id}', A_{id})$ topology - of linear forms $f \geq 0$ on A_{id} with $f(1) = 1$ and the topology to be chosen on S appears to be the topology induced by K. For $T \in A$, let $\varphi(T)$ be the function on S defined by $\varphi(T)f = f(T)$; each $\varphi(T)$ is a Borel function on the dense Borel set S in K, and $T \in A \longmapsto \varphi(T)$ is a positive ρ-norm preserving map from $A_{\mathbb{R}}$ onto $\varphi(A)$ (proposition 8.5). We thus find (proposition 8.7) that, to each $f = f^* \in A^\rho$, it corresponds a real Radon measure μ on K, such that one has $|\mu|(K-S) = 0$, $\varphi(A) \subset L^1(S, d\mu)$ and

$f(T) = \int_S \varphi(T)d\mu$: equivalently, given an integer $i \geq 0$, one has the unique decomposition $f = f_1 - f_2$ on A with elements $f_1 \geq 0$, $f_2 \geq 0$ in A^ρ and $\|f\|_{A_i} = \|f_1\|_{A_i} + \|f_2\|_{A_i}$ (for f in the predual P_A, one has $f_1, f_2 \in P_A$). This description is explicit for $B(\mathcal{D}, \mathcal{D})$ and other descriptions of the predual are indicated.

Chapter 9 introduces a notion of G-invariance. For an ultraweakly closed space A with condition II and cofinal abelian sequence A_i, we consider continuous linear forms f on A which are invariant under inner automorphisms associated to an abelian von Neumann algebra P containing all A_i^{-1}: linear forms invariant under inner automorphisms from $P^c = P' \cap A_{id}$ are called G-traces (this notation is due to the fact that one may choose $A_i = 1$ for a von Neumann algebra). For a G-invariant f, one can show that the above-mentioned decomposition $f = f_1 - f_2$ with $\|f\|_{A_i} = \|f_1\|_{A_i} + \|f_2\|_{A_i}$ for a given i must satisfy $\|f\|_{A_j} = \|f_1\|_{A_j} + \|f_2\|_{A_j}$ for all $j \geq 0$ (this fact was suggested by H. Araki) and that the modular one-parameter group σ_t^f of (A_{id}, f) induces a one-parameter group of isomorphisms of A. Extremal positive G-invariant linear forms on A correspond to characters of the abelian ultraweakly closed *-algebra $\mathcal{P} = \cup_{i\geq 0} PA_i \times A_i$ and a precise description of G-invariant states and G-traces is obtained when characters on A are normal.

Chapter 10 treats commutation theory as suggested by example 3. Due to the important role of $B(\mathcal{D}, \mathcal{D})$ - the analogue of the Banach algebra $\mathcal{L}(H)$ in W*-theory - one is tempted to calculate the commutant in $B(\mathcal{D}, \mathcal{D})$, as was done in [1]. Let us take an ultraweakly closed space A with condition II, and let M be the von Neumann algebra $M = A_{id}$. The striking fact is that a commutant for A does not necessarily lie in $B(\mathcal{D}, \mathcal{D})$, but has to be found in some $B(\mathcal{D}_1, \mathcal{D}_1)$, where $\mathcal{D}_1$ is a chosen dense domain 'affiliated' to the von Neumann algebra M'. The commutant $(A', \mathcal{D}_1)$ so obtained is an ultraweakly closed space with condition II in $B(\mathcal{D}_1, \mathcal{D}_1)$ (this last ultraweak topology refers to $\mathcal{D}_1 \hat{\otimes} \mathcal{D}_1$) and the double commutant property follows. It is shown that commutation between $A \equiv (A, \mathcal{D})$ and $A' \equiv (A', \mathcal{D}_1)$ is ensured by two *-algebras contained in A and A', respectively, and dense in these spaces (theorems 10.1 and 10.3). It is also proved that $\mathcal{D} \cap \mathcal{D}_1$ must contain an invariant linear set dense in each of the Fréchet spaces $\mathcal{D}$ and $\mathcal{D}_1$ which suffices for all commutations properties (proposition 3-13): this is the analogue of C_c^∞ in example 3.

We point out that a normal, faithful linear form $f \geq 0$ on A determines a privilegiate commutant for $\pi_f(A)$, and that this commutant is anti-isomorphic to $\pi_f(A)$. Finally, the commutant of a given *-algebra A_0 with condition II is identical to the commutant of its ultraweak closure A.

Chapter 11 introduces and studies strong and ultrastrong (also called σ-strong) topologies; these I have avoided previously in order to improve the readability of the book. These topologies behave as one expects and are the natural context for a generalized version of Kaplansky's density theorem.

VOCABULARY

Let A be a linear operator, not necessarily bounded, defined on a dense linear subset $\mathcal{D}$ of a Hilbert space H. The closure of A will be denoted by $\bar{A}$, the adjoint of A by A^+, and when Dom A^+ contains $\mathcal{D}$, the restriction of A^+ to $\mathcal{D}$ by A^*. We denote the set of integers by $\mathbb{N}$.

A set $\mathcal{A}$ of linear operators will be called a *-algebra on a domain $\mathcal{D}$ if the following conditions are satisfied:

1°/ $\mathcal{A}$ is symmetric, i.e., Dom $A^+ \supset \mathcal{D}$ and $A^* \in \mathcal{A}$ for any $A \in \mathcal{A}$.

2°/ $\mathcal{D}$ is invariant, i.e., $A\mathcal{D} \subset \mathcal{D}$ for any $A \in \mathcal{A}$.

3°/ $\mathcal{A}$ is an algebra, i.e., $A + B \in \mathcal{A}$, $AB \in \mathcal{A}$, $cA \in \mathcal{A}$ for any A, B $\in \mathcal{A}$, and any complex number c.

4°/ $\mathcal{A}$ is unital, i.e., $1 \in \mathcal{A}$.

An operator $A \in \mathcal{A}$ is positive (written $A \in \mathcal{A}^+$, or $A \geq 0$), if $(Ax, x) \geq 0$ for all $x \in \mathcal{D}$. The positive cone $\mathcal{A}^+$ is generating in $\mathcal{A}$ and defines the order structure $\leq$ of $\mathcal{A}$. For every symmetric operator T in $\mathcal{A}$, there exists a positive element A in $\mathcal{A}$ such that $-A \leq T \leq A$: this will be written simply as $|T| \leq A$. Moreover, we always assume that $\mathcal{A}$ is countably dominated, i.e., there exists a sequence A_n in $\mathcal{A}^+$, which is cofinal in $\mathcal{A}^+$.

The domain $\mathcal{D}$ of any *-algebra $\mathcal{A}$ will be equiped with the topology given by semi-norms $x \in \mathcal{D} \to \|Ax\|$, with A varying in $\mathcal{A}$. We note that $\mathcal{A}$ is countably dominated (by the sequence A_n) iff $\mathcal{D}$ is metric under this topology; in which case, semi-norms $x \in \mathcal{D} \to \|A_n x\|$ with $n \in \mathbb{N}$ suffice to define the topology of $\mathcal{D}$. The completion $\hat{\mathcal{D}}$ of $\mathcal{D}$ is $\hat{\mathcal{D}} = \cap_{A\in\mathcal{A}} \mathrm{Dom}(\bar{A})$ with

semi-norms $x \in \mathcal{D} \to \|\bar{A}x\|$, $A \in \mathcal{A}$, and our *-algebra $\mathcal{A}$ on the domain $\mathcal{D}$ can be considered as a *-algebra $\mathcal{A}$ on the domain $\hat{\mathcal{D}}$. When $\mathcal{D} = \hat{\mathcal{D}}$, we say that $\mathcal{D}$ is the natural domain of A. Before defining the topology ρ on $\mathcal{A}$ [11], which rests on the order structure of $\mathcal{A}$, we note that $(Ax, x) \geq 0$ for all $x \in \mathcal{D}$ is equivalent to $(\bar{A}x, x) \geq 0$ for all $x \in \hat{\mathcal{D}}$ (for some A in $\mathcal{A}$), and it follows that we may, and do, assume, without loss of generality, the equality $\mathcal{D} = \hat{\mathcal{D}}$.

Algebras handled in this paper will satisfy:

Condition I. There exists a cofinal sequence A_n in $\mathcal{A}^+$, such that $A_n \geq 1$, and $A_n^{-1} \in \mathcal{A}$ for every n.

Condition I clearly imply $A_n\mathcal{D} = \mathcal{D}$ for every integer n; however, one has to point out that it is not true, in general, that $A\mathcal{D} = \mathcal{D}$ for any $A \geq Id$, $A \in \mathcal{D}$. A weaker condition, often met in practice (in representations of enveloping algebras, for example) is

Condition I_0. There exists a cofinal sequence A_n in A^+ such that $A_n \geq Id$ and $A_n\mathcal{D} = \mathcal{D}$ for every $n \in \mathbb{N}$.

When condition I_0 is satisfied, the algebra A_1 generated by A and all A_n^{-1} (restricted to $\mathcal{D}$) is a $\star$-algebra on $\mathcal{D}$ satisfying condition I and the commutant A' of A and A_1 (A' is calculated inside $B(\mathcal{D}, \mathcal{D})$) are the same. It is seen in [1], theorem 3 that A' is always a $\star$-algebra on the domain $\mathcal{D}$, however, the natural domain D of A' is, in general, larger than $\mathcal{D}$.

At this point, there are two possibilities - a priori - for a second commutant A'' for A:

1°/ The commutant of $(A', \mathcal{D})$ calculated inside $B(\mathcal{D}, \mathcal{D})$ leads to the σ-weak closure A_1^σ - relative to $\mathcal{D} \hat{\otimes} \mathcal{D}$ - of the $\star$-algebra A_1. Note that A_1^σ is not, in general, a $\star$-algebra.

2°/ The commutant of (A', D) calculated inside $B(D, D)$ leads to an ultraweak closed $\star$-algebra - relatively to $D \hat{\otimes} D$ - with domain D. It often happens that this second commutant - always contained in A_1^σ - may consist of bounded operators only.
In fact, it is seen in Chapter 10 that commutation theory is more simple, and is a simple reality suggested by examples.

The complex conjugate $\mathcal{D}^-$ of $\mathcal{D}(\equiv \hat{\mathcal{D}})$ will be needed all through the book. As no confusion is possible in these notes, I have adopted a unified notation, i.e., $\mathcal{D} = \mathcal{D}^-$. I recall that the dual of the projective tensor product $\mathcal{D} \hat{\otimes} \mathcal{D}$ is the set $B(\mathcal{D}, \mathcal{D})$ of continuous sesquilinear forms on $\mathcal{D} \times \mathcal{D}$ (in the particular case when $\mathcal{D}$ is the Hilbert space H itself, $B(\mathcal{D}, \mathcal{D})$ coïncide with the Banach algebra $L(H)$ of bounded linear operators acting in H). Let B be a linear subset of $B(\mathcal{D}, \mathcal{D})$; the topologies on B induced by $\sigma(B(\mathcal{D}, \mathcal{D}), \mathcal{D} \otimes \mathcal{D})$ and $\sigma(B(\mathcal{D}, \mathcal{D}), \mathcal{D} \hat{\otimes} \mathcal{D})$ are called the weak and ultraweak (or σ-weak) topology of B. Thus, weak topology is defined by semi-norms $\beta \in B \to |\beta(x, y)|$, with x and y varying in $\mathcal{D}$, and σ-weak topology by semi-norms $\beta \in B \to |\sum_{i=1}^{\infty} \beta(x_i, y_i)|$, where $x = (x_i)_{i\in\mathbb{N}}$ and $y = (y_i)_{i\in\mathbb{N}}$ vary in the set of σ-convergent sequences (i.e., $\sum_{i=1}^{\infty} \|Ax_i\|^2 < +\infty$ for every $A \in A$ [11])

Any operator T of the $\star$-algebra A is identified to the continuous sesquilinear form $(x, y) \in \mathcal{D} \times \mathcal{D} \to (Tx, y)$, so that $\mathcal{A}$ is viewed as a vector subspace of $B(\mathcal{D}, \mathcal{D})$. In fact, the set $B^+(\mathcal{D}, \mathcal{D})$ of all β in $B(\mathcal{D}, \mathcal{D})$, such that $\beta(x, x) \geq 0$ for all x of $\mathcal{D}$, induces the order structure of $\mathcal{A}$, and defines the order structure $\leq$ in $B(\mathcal{D}, \mathcal{D})$: an important point is that any cofinal subset of $\mathcal{A}^+$ is automatically cofinal in $B^+(\mathcal{D}, \mathcal{D})$. The map $\beta \in B(\mathcal{D}, \mathcal{D}) \mapsto \beta^* \in B(\mathcal{D}, \mathcal{D})$, where, by definition, $\beta^*(x, y) = \overline{\beta(y, x)}$ for $x, y \in \mathcal{D}$ is an involution on $B(\mathcal{D}, \mathcal{D})$, and extends the involution of $\mathcal{A}$. Finally, for $\beta = \beta^*$ in $\mathcal{B}$, and $A \geq 0$ in $\mathcal{B}$, $|\beta| \leq A$ means $-A \leq \beta \leq A$.

For any subspace $\mathcal{B}$ of $B(\mathcal{D}, \mathcal{D})$ with $\mathcal{B} = \mathcal{B}_{\mathbb{R}} \oplus i\mathcal{B}_{\mathbb{R}}$, where $\mathcal{B}_{\mathbb{R}}$ is the set of all hermitian β in $\mathcal{B}$ (i.e., $\beta = \beta^*$), $\mathcal{B}$ containing a cofinal subset of the positive cone $\mathcal{A}^+$ of the given $*$-algebra $\mathcal{A}$, we define the ρ-topology as follows. For $A \in \mathcal{B}^+$, the set of all $T \in \mathcal{B}_{\mathbb{R}}$, such that $|T| \leq \lambda A$ for some $\lambda \in \mathbb{R}$, is denoted by $\mathcal{B}_A$ and the infimum of λ is denoted by $\rho_A(T)$ or $\|T\|_A$. Clearly, $\mathcal{B}_{\mathbb{R}} = \cup_{A\geq 0} \mathcal{B}_A$ and the locally convex inductive limit topology for the system of normed spaces $(\mathcal{B}_A, \rho_A)$ is the ρ-topology of $\mathcal{B}_{\mathbb{R}}$; the topology ρ of $\mathcal{B}$ is the direct topological sum $\mathcal{B}_{\mathbb{R}} \oplus i\mathcal{B}_{\mathbb{R}}$. Since $\mathcal{A}$ is countably dominated, we may choose a monotone increasing sequence A_n in $\mathcal{A}^+$ (or in $\mathcal{B}^+$) such that $\mathcal{B}_{\mathbb{R}} = \cup_n \mathcal{B}_{A_n}$, and $(\mathcal{B}_{\mathbb{R}}, \rho)$ coincide with the inductive limit of the sequence of normed space $(\mathcal{B}_{A_n}, \| \ \|_{A_n})$. We often identify $(\mathcal{B}_{\mathbb{R}}, \rho)$ with $(\mathcal{B}, \rho)$ in our notations. Let us recall that $(\mathcal{A}, \rho)$ is a separated $\mathcal{DF}$-space, that the strong dual of $(\mathcal{A}, \rho)$ is always a Fréchet space, and that any positive (resp. bijective positive) linear map from $\mathcal{B}$ into a space of the same type decreases (resp. preserve) the ρ-norms, hence is continuous (resp. bicontinuous).

Of course, one can consider any $*$-algebra $\mathcal{A}$, with domain $\mathcal{D}$, as a vector subspace of the set $B(\mathcal{D}, H)$ of continuous sesquilinear forms on the product $\mathcal{D} \times H$. Clearly, $B(\mathcal{D}, H)$ can be identified with the set of linear operators continuous from the Fréchet space $\mathcal{D}$ into the Hilbert space H, and $B(\mathcal{D}, H)$ is the dual of the projective tensor product $\mathcal{D} \hat{\otimes} H$; thus, in this context, it is possible to introduce weak topology, or σ-weak topology, with reference to $\mathcal{D} \otimes H$ or $\mathcal{D} \hat{\otimes} H$. This embedding of $\mathcal{A}$ into $B(\mathcal{D}, H)$ leads to a topology, called λ, which happens to be equal to ρ for a particular $*$-algebras (see [11] and paragraph 6). For any $*$-algebra $\mathcal{A}$ with domain $\mathcal{D}$ (or more generally, for a subspace of $B(\mathcal{D}, H)$), we introduce, for $T \in \mathcal{A}$, and $0 \neq A \in \mathcal{A}$, the quantity

$$\lambda_A(T) = \sup(\|Tx\| / \|Ax\|; \ x \in \mathcal{D})$$

with $a/0 = +\infty$ for $a > 0$, and the set $\mathcal{A}^A$ of all T in $\mathcal{A}$ being such that $\lambda_A(T)$ is finite. Each $\mathcal{A}^A$ is a linear space with

natural norm λ_A, and λ-topology is, by definition, the inductive limit topology of the system of normed spaces (A^A, λ_A), $A \in \mathcal{A}$. This topology is not related to the order structure of $\mathcal{A}$.
Subspaces $\mathcal{B}$ of $B(\mathcal{D}, \mathcal{D})$, of special interest, satisfy condition II.

Condition II. $\mathcal{B}$ is a linear subset of $B(\mathcal{D}, \mathcal{D})$, stable under involution and containing a cofinal subset A_n of $B^+(\mathcal{D}, \mathcal{D})$, with $A_n \geq 1$, $A_n(\mathcal{D}) = \mathcal{D}$ for every n, satisfying the following:

1°/ $\mathcal{B}_{id}$ is an involutive algebra of bounded operators containing 1.

2°/ $\beta \in B$ implies $A\ \beta\ A_n \in \mathcal{B}$ and $A_n^{-1}\ \beta\ A_n^{-1} \in \mathcal{B}$, where $(C\ \beta\ D)(x, y) = \beta(Dx, C^*y)$.

In general, elements of $\mathcal{B}_{id}$ do not send the domain $\mathcal{D}$ into itself. For any $B \in \mathcal{B}_{id}$ and any integer n, the continuous sesquilinear form $\beta : (x, y) \in \mathcal{D} \times \mathcal{D} \to (B\ A_n x, A_n y)$ belongs to $\mathcal{B}$, by 2°/, and will often be denoted by $BA_n \times A_n = \beta$.
The space $B(\mathcal{D}, \mathcal{D})$ satisfies condition II, and $B(\mathcal{D}, \mathcal{D})_{id}$ is the usual Banach algebra $\mathcal{L}(H)$, H being the Hilbert space completion of $\mathcal{D}$. The meaning of condition II comes from the following situation: for $*$-algebra $\mathcal{A}$ satisfying condition I, $\mathcal{A}_{id}$ is an involutive algebra of operators sending $\mathcal{D}$ into $\mathcal{D}$, the maps introduced in condition II 2°/ being the usual multiplication $T \in \mathcal{A} \to A_n\ T\ A_n \in \mathcal{A}$ and $T \in \mathcal{A} \to A_n^{-1}\ T\ A_n^{-1} \in \mathcal{A}$, and the σ-weak closure $\mathcal{A}^\sigma$ of $\mathcal{A}$ verify condition II, the set $(\mathcal{A}^\sigma)_{id}$ being exactly the von Neumann algebra generated by $\mathcal{A}_{id}$ [1].
For a $*$-algebra $\mathcal{A}$, conditions I and II are identical.
Finally, we will use also notation $C\ \beta D = \beta(D^* \cdot, C \cdot)$.
By a space $\mathcal{A}$ with condition II and natural domain $\mathcal{D}$, we mean that $\mathcal{D}$ is the natural domain of the $*$-algebra generated by all A_n, A_n^{-1}.

EXAMPLES AND OBSERVATIONS

1°/ Let X be a locally compact space, μ a positive measure on X with support X and H the Hilbert space $L^2(X, \mu)$. Let A_i be a sequence of μ-measurable functions on X with values in $[1, +\infty]$ such that $\mu(x \in X | A_i(x) = +\infty) = 0$. Let $L^\infty((A_i), X, \mu)$ be the set of all functions of the form gA_i with g moving in $L^\infty(X, \mu)$ and i moving in $\mathbb{N}$. The linear space $\mathcal{D} = \{g \in H/A_i g \in H\}$ for all $i \geq 0$ is necessarily dense in H and each $f \in L^\infty((A_i), X, \mu)$ is identified to the - closeable - linear operator T_f defined by $T_f g = fg$ for all $g \in \mathcal{D}$. Assuming that each A_i^2 is dominated a.e. by some A_j, we see that $L^\infty((A_i), X, \mu)$ is an abelian $\star$-algebra ultraweakly closed relative to $\mathcal{D} \hat{\otimes} \mathcal{D}$, with natural domain $\mathcal{D}$ and cofinal sequence $T_{A_i} \equiv A_i$, and the condition $A_i\mathcal{D} = \mathcal{D}$ is satisfied. Conversely, a general, ultraweakly closed abelian $\star$-algebra with condition I or II is isomorphic to such an object. In paragraph 6 it is seen that derivations on such function spaces are identically zero.

An abelian $\star$-algebra A with condition I always admits an injective Gelfand transformation and the set of characters of A appears to be a dense subset of the compact set of characters associated to the bounded part A_{id} of A. An example from spectral theory is as follows: let $A \geq \mathrm{Id}$ be a self-adjoint (one may take $A = 1 + x^2 - (\partial^2/\partial x^2)$ of 6°/ and A be the abelian $\star$-algebra generated by A and A^{-1} with natural domain $\mathcal{D} = \bigcap_{n \geq 0} \mathrm{Dom}\, \bar{A}^n$. The set Z of characters of A is homeomorphic to the spectrum of $\bar{A}^{-1} - \{0\}$, and the Gelfand transformation for A sends A^k onto the function $A_k(x) = x^k$ for $x \in Z$, $k \in \mathbb{Z}$.

2°/ Let G be a Lie group with Lie algebra $\mathfrak{g}$ and enveloping algebra $\mathfrak{A}(\mathfrak{g})$. Let $G \in G \longmapsto U_g$ be a continuous unitary representation of G in some Hilbert space H, and

$$\Delta = 1 - \sum_{i=1}^{n} dU(X_i)^2, \text{ where } (X_i)\ 1 \leq i \leq n \text{ is a basis for } \mathfrak{g}.$$

The natural domain (Fréchet space) $\mathcal{D}$ of the $\star$-algebra $dU(\mathfrak{A}(\mathfrak{g}))$ coïncides with the Fréchet space $C^\infty(U)$ of C^∞-vectors for U. The topology of $C^\infty(U)$ is also the topology of uniform convergence on compact sets of G of the function in $C^\infty(U)$ and each of its derivatives. Since any element T of $dU(\mathfrak{A}(\mathfrak{g}))$ is dominated by some Δ^n for suitable $n \geq 0$ - see [15], for example - (i.e., $|(Tx, x)| \leq M(\Delta^n x, x)$ for all $x \in \mathcal{D}$ and $M < +\infty$ independent of x), one has $\mathcal{D} = \cap_{n\geq 0} \text{Dom } \bar{\Delta}^n$, thus $\mathcal{D}$ may be seen as the Fréchet space $C^\infty(e^{it\Delta})$ of C^∞-vectors for the one-parameter group $t \mapsto e^{it\Delta}$. Every continuous sesquilinear form β on $\mathcal{D} \times \mathcal{D}$, i.e., $\beta \in B(\mathcal{D}, \mathcal{D})$, has a symbolic representation $\beta = B\Delta^k \times \Delta^k$ (i.e., $\beta(x, y) = (B\Delta^k x, \Delta^k y)$ for all $x, y \in \mathcal{D}$) with suitable $k \geq 0$ and B in $L(H)$. The $*$-algebra A generated by Δ^{-1} and $dU(\mathfrak{A}(\mathfrak{g}))$ is contained in $L(\mathcal{D})$ and condition I or II is satisfied for A; one has $A = \bigcup_{n\geq 0} A_{\Delta^n}$, with $\Delta^n \mathcal{D} = \mathcal{D}$ and $\mathcal{D} = \bigcap_{n\geq 0} \bar{\Delta}^{-n}(H)$. In fact, for every elliptic element $D \geq 1$ in $dU(\mathfrak{A}(\mathfrak{g}))$, such that all powers D^n ensures domination in the $*$-algebra generated by Δ, one similarly has $A = \cup_{n\geq 0} A_{D^n}$ with $D(\mathcal{D}) = \mathcal{D}$ and $\mathcal{D} = \cap_{n\geq 0} \bar{D}^{-n}(H)$; this follows from Nelson stinespring theorem and the ellipticity of D^n.

We note that an element $\beta \in B(\mathcal{D}, \mathcal{D})$ satisfying $\beta(U_g x, y) = \beta(x, U_g^* y)$ for all $x, y \in \mathcal{D}$, all $g \in G$, is automatically in the commutant A' of A; this fact is implicitly contained in [18].

3°/ We take, for U, the left regular representation of G defined by

$$(U_g f)(x) = f(g^{-1}.x)$$

for $g \in G$, $f \in L^2(G, dx) = H$, where dx is a fixed left invariant Haar measure on G. Similarly, we put

$$(V_g f)(x) = f(x \cdot g).$$

For $X \in \mathfrak{g}$, we identity $dU(X)$ with the right invariant differential operator

$$[dU(X)f](x) = \frac{d}{dt} f(\exp(-tX) \cdot x)\Big|_{t=0},$$

and, similarly, $dV(X)$ is the left invariant differential operator

$$[dV(X)f](x) = \frac{d}{dt} f(x \cdot \exp(tX))\Big|_{t=0}.$$

The enveloping algebra $dU(\mathfrak{A}(\mathfrak{g}))$ (resp. $dV(\mathfrak{A}(\mathfrak{g}))$) may be regarded as an algebra of right (resp. left) invariant differential operators on G, and it is also known that each of these algebras may be viewed as an algebra of scalar distributions on G. Assume that G is unimodular. Let $\mathcal{D}_u$ (resp. $\mathcal{D}_v$) be the space of C^∞-vectors for U (resp. V); one then has

$$\mathcal{D}_u = \{f \in C^\infty(G) \mid Xf \in H \quad \text{for all } X \in dU(\mathfrak{A}(\mathfrak{g}))\}$$

$$\mathcal{D}_v = \{f \in C^\infty(G) \mid Xf \in H \quad \text{for all } X \in dV(\mathfrak{A}(\mathfrak{g}))\}$$

and the anti-isometry $J: f \mapsto \check{f}$ where $\check{f}(x) = \bar{f}(x^{-1})$ exchanges $\mathcal{D}_u$ and $\mathcal{D}_v$. However, we note that $\mathcal{D}_u \neq \mathcal{D}_v$ in general, therefore the system of two perfectly commuting representations U and V shows that the natural domain for the commutant V(G) of U differs from the natural domain $\mathcal{D}_u$ of U (see paragraph 10 on commutation theorems). We note that $C_c^\infty(G) \subset \mathcal{D}_u \cap \mathcal{D}_v$ and this space is dense in each of the Fréchet spaces considered (compare with proposition 3.12) with

$$\Delta = dU(1 - \sum_{i=1}^{n} x_1^2), \; \nabla = dV(1 - \sum_{i=1}^{n} x_i^2)).$$

4°/ Let M be a von Neumann algebra acting in some Hilbert space H and $\Delta \geq \mathrm{Id}$ be a self-adjoint operator affiliated to M. Let $\mathcal{D} = \cap_{n\geq 0} \mathrm{Dom}\, \bar{\Delta}^n$ be the Fréchet space of differentiable vectors for Δ. The space $A \equiv \cup_{n\geq 0} M\Delta^n \times \Delta^n$ (i.e., the set of all sesquilinear forms β of description $\beta(x, y) = (B\Delta^n x, \Delta^n y)$ for all $x, y \in \mathcal{D}$) is an ultraweakly closed space - relative to $\mathcal{D} \hat{\otimes} \mathcal{D}$ - satisfying condition II and may be viewed as a prototype example: note that the bounded part of A is exactly M. This example is the neccessary concept for unbounded non-commutative integration, comparable to the concept of a von Neuman algebra in the bounded case. The biggest possible choice for A is the space $B(\mathcal{D}, \mathcal{D})$ of all continuous sesquilinear forms on $\mathcal{D} \times \mathcal{D}$ and, in fact, $B(\mathcal{D}, \mathcal{D}) = \cup_{n\geq 0} L(H)\Delta^n \times \Delta^n$.

In such A, there exists dense $\star$-algebras - with condition I and natural domain $\mathcal{D}$ - consisting of operators sending $\mathcal{D}$ into itself, so permitting a commutation theory in which commutants for A are spaces of continuous sesquilinear forms of the same kind (i.e., ultraweakly closed spaces with condition II). However, the commutant of A is not a unique object (except for its bounded part M') and depends on the choice of a suitable domain $\mathcal{D}_1 \neq \mathcal{D}$ (see paragraph 3, 10); this point is significant in example 3.

In such in A, one can perform the G.N.S. representation π_f associated to a positive linear form f on A. We show that the operator Δ is send into a self-adjoint operator $\pi_f(\Delta)$ and that the restriction of π_f to $M = A_{id}$ is exactly the usual G.N.S. representation associated to (M, f) with cyclic vector ζ_f. The image of each $\beta = B\Delta^k \times \Delta^k$ is the continuous sesquilinear form $\pi_f(\beta) = \pi_f(B)\pi(\Delta)^k \times \pi(\Delta)^k$ on $\mathcal{D}_f \times \mathcal{D}_f$,

where $\mathcal{D}_f = \cap_{n\geq 0} \text{Dom}\, \overline{\pi_f(\Delta)}^n$ and one has $f(\beta) = (\pi_f(\beta)\zeta_f, \zeta_f)$

as $\zeta_f \in D_f$. The map $T \in M \longmapsto j(T) = \pi_f(T)\zeta_f \in H_f$, where H_f is the Hilbert space associated to (M, f) extends uniquely as a map from A into a - dense - space of distributions on $\mathcal{D}_f$.

When Δ is central (i.e., $\Delta^{-1} \in M \cap M'$), the theory is closed to von Neumann algebra theory, since an ultraweakly closed space A with condition II must be a $\star$-algebra with natural domain $\mathcal{D}$ (in a separable Hilbert space, up to direct integrals diagonalizing $Z = M \cap M'$, we are reduced to M factor, hence Δ is a scalar operator $\neq 0$). <u>When Δ is not central, the situation is very different</u>, and it is not surprising that certain properties depend directly on the abelian $\star$-algebra generated by Δ and Δ^{-1}. Note also that the strong dual A^ρ of A and its predual P_A are automatically Fréchet spaces and A is endowed with a DF-space topology based on domination in A.

5°/ There is a closed relation between these developments and the general framework of quantum mechanics and field theory. Sesquilinear forms $\beta = \beta^*$ correspond to observables and the dynamics of the system is given and described by the operator Δ (one may also take $\Delta = 1 + H^*H \geq 1$, where H is the Hamiltonian). This imposes a set $\mathcal{D}$ of regular states, i.e., $\mathcal{D} = \cap_{n\geq 0} \text{Dom}\, \bar{\Delta}^n$ and we see that the iteration of a given β is not always a possible operation (the square β^2 of β is not defined for all β). Certain $\beta = \beta^*$ - but not all - may be described by self-adjoint operators with domain containing $\mathcal{D}$. If P is the family of projectors in the von Neumann algebra M, we also see that the set of all limits on $\mathcal{D} \times \mathcal{D}$ of sequences constructed with linear combinations in P is exactly A. The evolution in time of a given β is the map $t \in \mathbb{R} \longmapsto \beta_t \equiv \beta(e^{it\Delta}\cdot, e^{it\Delta}\cdot) \in A$ and for β bounded - i.e., $\beta = B \in L(H)$ - or $\beta \in L(\mathcal{D})$, we find the usual condition $\beta_t = e^{-it\Delta}\beta e^{it\Delta}$.

The spectral nature of Δ imposes limitations on the space state of A: for example, we find that, for a Δ^{-1} compact operator in H, the strong dual of A or of

$B(\mathcal{D}, \mathcal{D}) = \cup_{n\geq 0} L(H)\Delta^n \times \Delta^n$ coïncide as Fréchet space with

its predual.
Equivalently, every state on A is given by a trace class operator sending H into $\mathcal{D}$ and conversely: it is known that this property fails for $L(H)$. The eternal 'moment problem' appears connected with the existence of positive linear maps having prescribed values on all Δ^n (proposition 5.5). One notes that sesquilinear forms are necessary mathematical objects for field theory.

6°/ A classical example in analysis is the Schrödinger representation U of the Heisenberg group A_n; we take n = 1 for simplicity. The Fréchet space of C^∞-vectors for U is the Schwartz space $S = S(\mathbb{R})$. Denoting by x_1, x_2, x_3 a suitable basis of $\mathfrak{g}$, the infinitesimal representation $dU(\mathfrak{g})$ acts as

$$dU(X_1) = \frac{d}{dx}\ ; \qquad dU(X_2) = ix; \qquad dU(X_3) = i\ \mathrm{Id}$$

on $S \subset L^2(\mathbb{R}) = H$. Putting, as usual, $p = (1/i)(d/dx)$ and $q = x$, we know that p, q are essentially self-adjoint on

$$S = \cap_{n\geq 0} \overline{\mathrm{Dom}(p^2 + q^2 + 1)^n} = \cap_{n\geq 0} (p^2 + q^2 + 1)^{-n}(H).$$

We now introduce some notation: given a self-adjoint operator T in an Hilbert space H, the Paley-Wiener space for $T = \int_{-\infty}^{\infty} \lambda\ d\ p_\lambda$ is the space $\mathcal{D}_{pw}(T) = \cup_{n\geq 0} E_n^T(H)$, where

$E_n^T = \int_{-n}^{n} dp(\lambda)$, and $C^\infty(T) = \cap_{n\geq 0} \mathrm{Dom}\ \bar{T}^n$ is the space of C^∞-vectors for $t \longmapsto e^{itT}$ (note that $\mathcal{D}_{pw}(T) = \mathcal{D}_{pw}(1+T^2)$ and $C^\infty(T) = C^\infty(1+T^2)$). Taking T = q, we find that $\mathcal{D}_{pw}(q) \cap S = C_c^\infty$ is the space of C^∞ functions with compact support.
Taking T = p we find that $\mathcal{D}_{pw}(p) \cap S = Z$ is the space of rapidly-decreasing functions with compact spectra. One has $F(C_c^\infty) = Z$ (F stands for the Fourier transformation) and the topology considered on Z is the image by F of the usual topology of C_c^∞ (i.e., the inductive limit as K compact increases of the Fréchet spaces C_K^∞). We note that the topologies induced by the Fréchet space S on (the DF-spaces) C_c^∞ and on Z are less fine than topologies indicated. However, these usual topologies of C_c^∞ and of Z can be constructed via the Paley-Wiener space and the topology of S. Indeed, let

us write $E = C_c^\infty$ or $Z \leftrightarrow$ with their usual topologies, and put $T = p$ or q. As vector spaces, one has $E = \bigcup_n E_n$ where $E_n \equiv E_n^T(H) \cap S$, and noting that E is closed in the Fréchet space S, we find that the strict inductive limit of the sequence of Fréchet Spaces E_n makes sense and is exactly the topology given on E. Indeed, for $E = C_c^\infty$, E_n is the Fréchet space $C_{K_n}^\infty$ with $K_n = [-n, n]$ and for $E = Z$, it suffices to note that F exchanges $E_n^{(p)}$ and $E_n^{(q)}$ and is a topological isomorphism from S onto itself.
Therefore, for a space A with condition I or II and natural domain $\mathcal{D}$, we will refer to <u>the dual space $\mathcal{D}'$</u> (or anti-dual space $\mathcal{D}'$ - same notation) <u>as the space of distributions on $\mathcal{D}$</u>. We also easily see that each $\beta = B\Delta^k \times \Delta^k \in B(\mathcal{D}, \mathcal{D})$ induces a continuous linear map $T_\beta = \bar{\Delta}^k B \Delta^k$ from the DF-space $\mathcal{D}_{pw}(\Delta)$ into H (the converse fact is false), taking on $\mathcal{D}_{pw}(\Delta)$ the inductive limit topology of the sequence of Hilbert spaces $E_n^\Delta(H)$.

7°/ Many situations in mathematics or physics depend on the action of a given Lie group G on a smooth manifold V (e.g., G is the Poincaré group and $V = \mathbb{R}^4$) or in a given Hilbert space H (e.g. $H = L^2(V, d\mu)$ where $d\mu$ is a G-invarinat measure). Writing $(g, x) \in G \times V \longmapsto g \cdot x \in V$ or $g \to U_g \in L(H)$ for these actions, let $X_1, \ldots, X_n$ be a basis of the Lie algebra $\mathfrak{g}$ of G viewed either as differential operators on V or as skew-symmetric operators in H. Spaces $A_{(\Delta)} = \bigcup_k M \Delta^k \times \Delta^k$ introduced in example 4°/ correspond to the one-parameter action $t \longmapsto e^{it\Delta}$ (i.e., $G = \mathbb{R}$). They often suffice for the acquisition of information on a general space A with condition II containing all $(1 - \bar{X}_j^2)^{-1}$ for $1 \le j \le n$. For example, noting that C^∞-vectors for $g \longmapsto U_g$ are exactly C^∞-vectors for all parameters groups $t \to e^{it\bar{X}_j}$ by Goodman theorem, we find that σ-weakly continuous linear forms f on A are those f defined and σ-weakly continuous on all $A_j \equiv \bigcup_{k \ge 0} M(1-\bar{x}_j^2)^k \times (1-\bar{x}_j^2)^k$ for $1 \le j \le n$, where $M = A_{id}$. Using theorem 9.1, we thus find that a G-invariant form $f = f^*$ has a unique minimal decomposition in positive G-invariant f_1, f_2 on A with $f = f_1 - f_2$.

CHAPTER 1: MAIN SPACES

We will keep to the notation of the vocabulary section. Objects of main interest are mentionned in lemma 1.1 and theorem 1.1.

Lemma 1.1. Let A_i be a cofinal sequence in $B^+(\mathcal{D}, \mathcal{D})$, with $A_i \geq Id$, $A_0 = Id$, and $A_i\mathcal{D} = \mathcal{D}$ for every $i \in \mathbb{N}$.

1°/ Let $\mathcal{C}$ and $\mathcal{E}$ be subspaces of $B(\mathcal{D}, \mathcal{D})$, stable under involution containing 1 and stable under operations $\beta \to A_n \beta A_n$ and $\beta \to A_n^{-1} \beta A_n^{-1}$, for every $n \geq 0$, with $\mathcal{C} \subset \mathcal{E}$. Then $\mathcal{C}_{id} = \mathcal{E}_{id}$ is equivalent to $\mathcal{C} = \mathcal{E}$.

2°/ For any n, one has $\varphi_n(\mathcal{C}_{id}) = \mathcal{C}_{A_n^2}$, φ_n being the map $\beta \to A_n \beta A_n$.

Theorem 1.1. Let M be a von Neumann algebra containing the A_n^{-1}, for all $n \in \mathbb{N}$. Let $\mathcal{B}$ be the subspace of $B(\mathcal{D}, \mathcal{D})$ equal to $\bigcup_n M A_n \times A_n$ (i.e., the set of forms $(x, y) \in \mathcal{D} \times \mathcal{D} \to (BA_nx, A_ny)$, B varying in M and n in $\mathbb{N}$). Then $\mathcal{B}$ is σ-weakly closed in $B(\mathcal{D}, \mathcal{D})$ and satisfies condition II. For every n, one has $\mathcal{B}_{A_n^2} = M A_n \times A_n$, and, in particular, $\mathcal{B}_{id} = M$.

It follows that the σ-weak closure $\mathcal{B}^\sigma$ of a space $\mathcal{B}$ satisfying condition II is $\bigcup_n M A_n \times A_n$, M being the von Neumann algebra generated by $\mathcal{B}_{id}$.
Property 2°/ of lemma 1.1 is a lift proces in $\mathcal{B}$ which often reduces many properties of $\mathcal{B}$ into properties in $\mathcal{B}_{id}$. Note that $\mathcal{B}_{id}$ is an involutive algebra of bounded operators acting in the Hilbert space H (completion of $\mathcal{D}$), and that φ_i is isometric form $(\mathcal{B}_{id}, \| \; \|_{id})$ onto $(\mathcal{B}_{A_i^2}, \| \; \|_{A_i^2})$, hence inducing a homeomorphism from $(\mathcal{B}, \rho)$ onto itself (with inverse $\varphi_{-i} : \beta \mapsto \varphi_{-i}(\beta) = A_i^{-1} \beta A_i^{-1}$)

Proofs. The first assertion of the lemma is straightforward. Now, for $B \in \mathcal{C}_{id}$, $n \in \mathbb{N}$, one has, for x in $\mathcal{D}$,

$$|(B A_n x, A_i x)| \leq \|B\| (A_n^2 x, x),$$

hence $\varphi(\mathcal{C}_{id}) \subset \mathcal{C}_{A_n^2}$. Conversely, let $\beta \in \mathcal{C}_{A_n^2}$; since A_i^2 is an order-unit of $\mathcal{C}_{A_i^2}$, we can assume $0 \leq \beta \leq A_i^2$. The form $(x, y) \in \mathcal{D} \times \mathcal{D} \to \beta(x, y)$ is hermitian, thus, by the Cauchy-Schwartz inequality, one has

$$|(x, y)| \leq \|A_i x\| \|A_i y\|$$

for x, y in $\mathcal{D}$. It follows that existence of a bounded operator $B \in L(H)$, with $B = B^*$, such that $\beta(x, y) = (B A_i x, A_i y)$. One has $B = A_i^{-1} \beta A_i^{-1}$, thus $B \in \mathcal{C}$ and being bounded, $B \in \mathcal{C}_{id}$. Hence, $\varphi(\mathcal{C}_{id}) = \mathcal{C}_{A_n^2}$.

Let β be in $\mathcal{B}$ (of theorem 1.1); then $\beta = B A_i \times A_i$ for some integer i, and some $B \in M$. One has $A_n \beta A_n = B A_i A_n \times A_i A_n$. But, one can find an integer j and a finite constant M such that $\|A_i A_n x\| \leq M \|A_j x\|$ for all x in $\mathcal{D}$. Clearly, $A_i A_n A_j^{-1}$ is bounded and belongs to M, as well as its adjoint, which coïncide on $\mathcal{D}$ with $A_j^{-1} A_n A_i$: it follows that $A_n \beta A_n \in \mathcal{B}$. The proof is similar for $A_n^{-1} \beta A_n^{-1}$. One has $M \subset \mathcal{B}_{id}$ and, to get equality, it is sufficient to show that elements $T = T^*$ of M' commute with $\mathcal{B}_{id}$. By definition, T commutes with elements of M, in particular with $\bar{A}_i^{-1}$, hence $T\ (\mathrm{Dom}\ \bar{A}_i) \subset \mathrm{Dom}(\bar{A}_i)$ thus $T\mathcal{D} \subset \mathcal{D}$. By standard arguments, we get $TA_i x = A_i Tx$ for $x \in \mathcal{D}$. Now, for $\beta = BA_n \times A_n \in \mathcal{B}$ and $x, y \in \mathcal{D}$, one has $\beta(Tx, y) = \beta(x, Ty)$; indeed, $\beta(Tx, y) = (BA_n Tx, A_n y) = (BTA_n x, A_n y) = (TBA_n x, A_n y) = \beta(x, Ty)$. If $\beta \in \mathcal{B}_{id}$, then β corresponds to some bounded operator B_1 of $L(H)$ (i.e., $\beta(x, y) = (B_1 x, y)$ for x, y in $\mathcal{D}$), and our relations imply $B_1 \in (M')' = M$. Finally, $\mathcal{B}_{id} = M$. By [1], $\mathcal{B}$ is σ-weakly closed and the preceding lemma implies $\mathcal{B}_{A_n^2} = M A_n \times A_n$.

This proves our assertions.

Due to the fundamental role of $L(\mathcal{D})$ in this paper, we will now formulate the known

Definition 1.1. The set $L(\mathcal{D})$ of operator S, such that Dom $S \supset \mathcal{D}$, Dom $S^+ \supset \mathcal{D}$ satisfying $S\mathcal{D} \subset \mathcal{D}$, $S^*\mathcal{D} \subset \mathcal{D}$ is a $*$-algebra, with condition I, natural domain $\mathcal{D}$ and cofinal sequence A_i. One has $L(\mathcal{D}) \subset B(\mathcal{D}, \mathcal{D})$.

We often use

Lemma 1.2. 1°/ Let $\gamma \in B(\mathcal{D}, \mathcal{D})$ and $A \in L(\mathcal{D})$, such that $\gamma(Ax, y) = \gamma(x, A^*y)$ for all x, y in a dense linear subset $\mathcal{D}_0$ of the Fréchet space $\mathcal{D}$. Then $\gamma(Ax, y) = \gamma(x, A^*y)$ for all $x, y \in \mathcal{D}$.

2°/ For a given $\gamma = \gamma^* \in B(\mathcal{D}, \mathcal{D})$, the set $C_\gamma = \{A \in L(\mathcal{D}) \mid \gamma(Ax, y) = \gamma(x, A^*y) \text{ for all } x, y \in \mathcal{D}\}$ is an involutive subalgebra of $L(\mathcal{D})$. If $A \in C_\gamma$ with $A\mathcal{D} = \mathcal{D}$, $A^{-1} \in C_\gamma$.

3°/ For a given $A = A^* \in L(\mathcal{D})$, the set $C_A = \{\gamma \in B(\mathcal{D}, \mathcal{D}) \mid \gamma(Ax, y) = \gamma(x, A^*y) \text{ for all } x, y \in \mathcal{D}\}$ is a weakly and ultraweakly closed linear subset of $B(\mathcal{D}, \mathcal{D})$, stable under involution.

Proof. For 1°/, let $C \in L(H)$ and $i \geq 0$, such that $\gamma = CA_i \times A_i$, and $x, y \in \mathcal{D}$. We choose j in $\mathbb{N}$ and $M < +\infty$, such that $(A_iA)^* A_iA \leq M^2A_j^2$, $(A_iA^*)^* A_iA^* \leq M^2 A_j^2$ and $A_i^2 \leq M A_j^2$. Introducing a sequence x_n (resp. y_n) in $\mathcal{D}_0$, such that $\|A_j(x_n-x)\|$ (resp. $\|A_j(y_n-y)\|$) tends to zero, the relation $(CA_iAx_n, A_iy_n) = (CA_ix_n, A_iA^*y_n)$ yields the result as $n \to \infty$, due to

$$\|C(A_iA)(x_n-x)\| \leq M\|C\| \, \|A_j(x_n-x)\|$$

and

$$\|A_iA^*(y_n-y)\| \leq M\|A_j(y_n-y)\|; \qquad \|A_i(y_n-y)\| \leq M\|A_j(y_n-y)\|.$$

For 2°/, let $C = C^* \in L(H)$, $i \geq 0$ such that $\gamma = CA_i \times A_i$ and $x, y \in \mathcal{D}$. For $A \in C_\gamma$, the formula

$$(C A_i Ax, A_iy) = (C A_ix, A_iA^*y)$$

implies that

$$(C\,A_i\,A^*x,\ A_iy) = (A_i\,A^*x,\ C\,A_iy) = \overline{(C\,A_iy,\ A_iA^*x)}$$

$$= \overline{(C\,A_i\,Ay,\ A_ix)} = (C\,A_ix,\ A_iAy),$$

thus $A^* \in C_\gamma$. Taking B_1 and B_2 in C_γ, one clearly has $B_1 + B_2 \in C_\gamma$, $\lambda B_1 \in C_\gamma$ for complex λ. Formula $CA_iB_1 x A_i = CA_i \times A_iB_1^*$ (resp. $CA_iB_2 \times A_i = CA_i \times A_iB_2^*$) implies that $CA_iB_1B_2 \times A_i = CA_iB_2 \times A_iB_1^*$ (resp. $CA_iB_2 \times A_iB_1^* = CA_i \times A_iB_2^*B_1^*$) as seen by x replaced by B_2x (resp. y replaced by B_1^*y), hence $B_1B_2 \in C_\gamma$. When $A \in C_\gamma$, with $A\mathcal{D} = \mathcal{D}$ replacing x by $A^{-1}x$ and y by $(A^*)^{-1}y$, we obtain

$$(CA_ix,\ A_i(A^{-1})^*y = (CA_i\,A^{-1}x,\ A_iy),$$

hence $A^{-1} \in C_\gamma$. For 3°/, let I be an index set and $\gamma_\alpha \in C_A$, $\alpha \in I$ tending to $\gamma \in B(\mathcal{D}, \mathcal{D})$ as $\alpha \to \infty$. As $\gamma_\alpha(u, v) \to \gamma(u, v)$ for all $u, v \in \mathcal{D}$, taking $u = Ax$ and $v = y$ (resp. $u = x$ and $v = A^*y$), we thus prove our assertions.

We now turn to a slight modification of theorem 3 in [1], which deals with a $*$-algebra given in $L(\mathcal{D})$. This will bring to light those spaces A of interest in this paper (see theorem 3.1 and remark 1.1). We start here with an involutive algebra M_0 of bounded operators (acting in the Hilbert space H completion of $\mathcal{D}$), M_0 containing all A_i^{-1}, $i \geq 0$ with, as usual, $A_0 = \mathrm{Id}$, $A_n\mathcal{D} = \mathcal{D}$ for $n \geq 0$, and $\mathcal{D} = \cap_{n\geq 0} A_n^{-1}(H)$. We note that M_0 is not necessarily contained in $L(\mathcal{D})$. Let M be the von Neumann algebra generated by M_0 with commutant M'; clearly, $M'\,\mathcal{D} \subset \mathcal{D}$ (i.e., $M' \subset L(\mathcal{D})$). We put $\mathcal{B}_0 = \cup_{n\geq 0} M_0A_n \times A_n$, $\mathcal{B} = \cup_{n\geq 0} M\,A_n \times A_n$ and

$$\mathcal{B}_1 = \{\beta \in B(\mathcal{D}, \mathcal{D}) \mid \beta(Bx, y) = \beta(x, B^*y) \quad \text{for all } x, y \in \mathcal{D},\ B \in M'\};$$

$$\mathcal{B}' = \{S \in L(\mathcal{D}) \mid (ASx, y) = (Ax, S^*y) \quad \text{for all } x, y \in \mathcal{D},\ A \in M_0\}$$

The fact that $\mathcal{B}'$ is automatically contained in $L(\mathcal{D})$ rests on

Lemma 1.3. Let $\beta \in B(\mathcal{D}, \mathcal{D})$, such that $\beta(Ax, y) = \beta(x, A^*y)$ for all x, y $\in \mathcal{D}$ and $A \in M_0 \cap L(\mathcal{D})$. Then, $\beta \in L(\mathcal{D})$.
Consistent with theorem 3 in [1] is

Theorem 1.2. 1°/ One has $\mathcal{B}_1 = \mathcal{B} = \{\beta \in B(\mathcal{D}, \mathcal{D}) \mid \beta(Sx, y) = \beta(x, S^*y),\ x, y \in \mathcal{D} \text{ and } S \in \mathcal{B}'\}$. Spaces M_0, $\mathcal{B}_0$ are σ-weakly dense in $\mathcal{B}$ relative to $\mathcal{D} \hat{\otimes} \mathcal{D}$.

2°/ $\mathcal{B}'$ is a ⋆-algebra with condition I and cofinal central sequence $[A_n]_Z$ (see [1] or lemma 10.1 for the meaning), has natural domain $D \supset \mathcal{D}$. One has $\mathcal{B}' = \{S \in L(\mathcal{D}) \mid (ASx, y) = (Ax, S^*y),\ x, y \in \mathcal{D},\ A \in M\}$ and $\mathcal{B}'$ is algebraically generated by $\mathcal{B}'_{id} = M'$ and the $[A_n]_Z$, $n \geq 0$. The space $\mathcal{B}'$ is σ-weakly closed relatively to $\mathcal{D} \hat{\otimes} \mathcal{D}$ and $\mathcal{D} \hat{\otimes} \mathcal{D}$.

3°/ $A \equiv L(\mathcal{D}) \cap \mathcal{B}$ is a ⋆-algebra with condition I, cofinal sequence A_i and natural domain $\mathcal{D}$, and is exactly the set of elements in $L(\mathcal{D})$ which commute (in the usual sense on $\mathcal{D}$) with $\mathcal{B}'$. The ultraweak closure of A in $B(\mathcal{D}, \mathcal{D})$ is contained in $\mathcal{B}$, and will coïncide with $\mathcal{B}$ iff the von Neumann algebra M contains an involutive subalgebra in $M \cap L(\mathcal{D})$ which generates M as a von Neumann algebra.

Density of M (resp. of M') in $\mathcal{B}$ (resp. $\mathcal{B}'$) will be studied in more detail in the next paragraph. Since other commutants for $\mathcal{B}$ will be considered later, we will be led sometimes to write $\mathcal{B}' = (\mathcal{B}', \mathcal{D}) = (\mathcal{B}', D)$, i.e., $\mathcal{B}' \subset B(\mathcal{D}, \mathcal{D})$.
In the course of this paper, we shall see that it is necessary to replace the notation of $\mathcal{D}$ by that of $\mathcal{D}_{(A_i)}$, recalling that $\mathcal{D}$ is constructed and depends only on a chosen sequence A_i.
To make these notes readable we shall keep the notation $\mathcal{D}$ until paragraph 10.
Ultraweakly closed spaces with condition II are very natural objects (a direct extension of a von Neumann algebra) in which the ⋆-algebra $\mathcal{B} \cap L(\mathcal{D})$ plays a fundamental role. It will be seen in paragraph 3 that this ⋆-algebra is ultraweakly dense in $\mathcal{B}$ if the cofinal sequence A_i is abelian, thus explaining the technicalities of this work (the reader is also referred to remark 1.1.1°/), i.e., $\mathcal{B} \cap L(\mathcal{D})$ suffices to describe $\mathcal{B}$.

Proof of lemma 1.3. We may assume $A_i \leq A_j$ for $i \leq j$ and $\beta = \beta^*$. Let $C^* = C \in L(H)$ and $i \geq 0$, such that $\beta = C\,A_i \times A_i$. Note that C depends on the choice of i, i.e., $C = C_i$. In formula $C\,A_iA \times A_i = C\,A_i \times A_iA^*$, taking $A = A_i^{-1}$ we find $C \times A_i = C\,A_i \times Id$ on $\mathcal{D} \times \mathcal{D}$ leading to $C\,\mathcal{D} \subset Dom(\bar{A}_i)$ since A_i is essentially self-adjoint on $\mathcal{D}$. Thus, β corresponds to the linear operator $\bar{A}_i \; C \; A_i = \beta$ which sends $\mathcal{D}$ into Dom $(\bar{A}_i)$. Since for every $j \geq i$ there exists $C_j \in L(H)$, such that $\beta = C_j \; A_j \times A_j$, we see that β corresponds to an operator β sending $\mathcal{D}$ into $\cap_{j\geq i}$ Dom $(\bar{A}_j) = \mathcal{D}$ and the commutation formula becomes $(\beta \, Ax, y) = (\beta x, A^* y)$.

Proof of theorem 1.2. Since M_0 is Hilbert weakly dense in M, it is clear that

$$\mathcal{B}' = \{S \in L(\mathcal{D}) \mid (ASx, y) = (Ax, S^* y) \quad x, y \in \mathcal{D}, A \in M\}.$$

For every $\beta \in B(\mathcal{D}, \mathcal{D})$, there exists $B \in L(H)$, and integer n such that $\beta = BA_n \times A_n$. Let us establish first assertion. The proof of theorem 1.1 shows that $\mathcal{B} \subset \mathcal{B}_1$. Now let $\beta = BA_n \times A_n$ be in $\mathcal{B}_1$; one has, for $x, y \in \mathcal{D}$, $A \in M'$, $(BA_n \, Ax, A_n y) = (BA_n x, A_n A^* y)$. Since $A \, \bar{A}_n^{-1} = \bar{A}_n^{-1} \, A$ in the Banach algebra $L(H)$, and since $A(\mathcal{D}) \subset \mathcal{D}$, one has $A \, A_n x = A_n Ax$ on $\mathcal{D}$, hence, taking $u = A_n x$, $v = A_n y$, we get the relation $(B \, Au, v) = (Bu, A^* v)$, for all $u, v \in \mathcal{D}$, which holds by continuity for u, v in H, showing that $B \in (M')' = M$, which means that $\beta \in \mathcal{B}$. Finally, $\mathcal{B} = \mathcal{B}_1$. The density of $\mathcal{B}_0$ into $\mathcal{B}$ remains to show that, given a σ-convergent sequence $x = (x_i)_{i \in \mathbb{N}}$ in $\mathcal{D}$, one can find $\beta_0 \in \mathcal{B}_0$ such that $\left| \sum_{i=1}^{\infty} (\beta - \beta_0)(x_i, x_i) \right| \leq \varepsilon$, $\varepsilon > 0$ being given, with $\beta = B \, A_n \times A_n$, and $B \in M$. Since M_0 is Hilbert σ-weakly dense in M, and since the sequence $y = (y_i)_{i \in \mathbb{N}}$, with $y_i = A_n x_i$ is such that $\sum_{i=1}^{\infty} \|y_i\|^2 < +\infty$, there exists B_0 in M_0 satisfying

$$\left| \sum_{i=1}^{\infty} ((B - B_0) y_i, y_i) \right| \leq \varepsilon,$$

thus

$$\left| \sum_{i=1}^{\infty} ((B - B_0) A_n x_i, A_n x_i) \right| \leq \varepsilon,$$

which shows that $\beta_0 = B_0 \; A_n \times A_n$ is admitting. We now denote by $\mathcal{B}_2$ the set

$$\mathcal{B}_2 = \{\beta \in B(\mathcal{D}, \mathcal{D}) \mid \beta(Sx, y) = \beta(x, S^* y), \quad x, y \in \mathcal{D}, \; S \in \mathcal{B}'\}.$$

From $M' \subset \mathcal{B}'$, it is clear that $\mathcal{B}_2 \subset \mathcal{B}$. Now let $\beta \subset \mathcal{B}$ with $\beta = BA_n \times A_n$, $B \in M$. One has, for $S \in \mathcal{B}'$, $S = S^*$, $x, y \in \mathcal{D}$ $(BSx, y) = (Bx, Sy)$, in particular $(\bar{A}_n^{-1} Sx, y) = (A_n^{-1}x, Sy)$, hence $A_n^{-1}S = SA_n^{-1}$ on $\mathcal{D}$, implying $A_nS = S A_n$ on $\mathcal{D}$ (since $S\mathcal{D} \subset \mathcal{D}$ and $A_n\mathcal{D} = \mathcal{D}$). Thus, for $S = S^*$ in $\mathcal{B}'$, x, y in $\mathcal{D}$,

$$\beta(Sx, y) = (B A_n Sx, A_ny) = (B S A_nx, A_ny)$$

and

$$\beta(x, Sy) = (B A_nx, A_nSy) = (B A_nx, S A_ny).$$

Replacing x by A_nx, and y by A_ny in relation $(B Sx, y) = (Bx, Sy)$ with $B \in M$, it follows that $\beta(x, Sy) = \beta(Sx, y)$, hence $\beta \in \mathcal{B}_2$, since any element S of $\mathcal{B}'$ is a linear combination of hermitian elements in $\mathcal{B}'$. By lemma 1.2, $\mathcal{B}'$ is a $*$-algebra and equality $\mathcal{B}'_{id} = M'$ is clear. From proposition 2.1 (of paragraph 2), M is ultraweakly dense in $\mathcal{B}$, and hence in $\mathcal{B}_0$ by 1°/. But, M_0 is Hilbert ultraweakly dense in M, and it follows that M_0 is dense in M relative to $\mathcal{D} \hat{\otimes} \mathcal{D}$, proving 1°/.

Let us now recall that any $A \in L(\mathcal{D})$ is identified to the sesquilinear form $A:(x, y) \in \mathcal{D} \times \mathcal{D} \to (Ax, y)$. Since $L(\mathcal{D}) \subset B(\mathcal{D}, \mathcal{D})$, we get

$$\begin{aligned} \mathcal{B} \cap L(\mathcal{D}) &= \{A \in L(\mathcal{D}) \mid A(Sx, y) = A(x, S^*y),\\ &\qquad x, y \in \mathcal{D}, \ S \in \mathcal{B}'\}\\ &= \{A \in L(\mathcal{D}) \mid AS = SA \text{ on } \mathcal{D}, \ S \in \mathcal{B}'\}, \end{aligned}$$

therefore A is a $*$-algebra, satisfying condition I since it contains A_n and A_n^{-1}, for $n \in \mathbb{N}$. The second commutant of A in $B(\mathcal{D}, \mathcal{D})$ is the σ-weak closure A^σ of A relatively to $\mathcal{D} \hat{\otimes} \mathcal{D}$, by [1], hence $A^\sigma \subset \mathcal{B}$. When $A^\sigma = \mathcal{B}$, it follows from lemma 8.5 [1] that A_{id} is an involutive algebra of bounded operators, strongly dense in M. Conversely, let us assume the existence of an involutive algebra $N \subset L(\mathcal{D}) \cap M$, strongly dense in M. One can suppose that $A_n^{-1} \in N$ for every n, since the algebra generated by N and the A_n^{-1} has the same properties. The $*$-algebra C generated by N and operators A_n, with $n \in \mathbb{N}$ is clearly dominated by operators A_n, hence admits $\mathcal{D}$ as natural domain. Of course, $N \subset \mathcal{C}_{id} \subset M$, since $\mathcal{C} \subset \mathcal{B} \cap \mathcal{L}(\mathcal{D})$. It is immediately seen that $\mathcal{C}$ satisfy condition I, thus the σ-weak closure of $\mathcal{C}$, relative to $\mathcal{D} \hat{\otimes} \mathcal{D}$, is the space $\cup_{n\geq 0}(\mathcal{C}_{id})'' \, A_n \times A_n = \mathcal{B}$.

We will now achieve the proof of the second assertion. Let $S \in \mathcal{B}'$; since $\mathcal{A}_{id} \subset M$, we get, for $x, y \in \mathcal{D}$, $A \in \mathcal{A}_{id}$, $(ASx, y) = (Ax, S^*y)$, which lead to $ASx = SAx$ for $x \in \mathcal{D}$ (due to $A\,\mathcal{D} \subset \mathcal{D}$). For any A in $\mathcal{A}$, we can find an integer i such that $A_i^{-1} A A_i^{-1} \in \mathcal{A}_{id}$, hence, with the help of relation $A_i^{-1} S = S A_i^{-1}$ on $\mathcal{D}$, we get, for $x, y \in \mathcal{D}$

$$(A_i^{-1} A A_i^{-1} S x, y) = (S A_i^{-1} A A_i^{-1} x, y)$$

and, putting $\xi = A_i^{-1} x$, $\eta = A_i^{-1} y$, we obtain

$$S(\xi, A^* \eta) = S(A\xi, \eta)$$

for all $\xi, \eta \in \mathcal{D}$, thus $S \in \mathcal{A}' \subset \mathcal{L}(\mathcal{D})$ by lemma 1.3. In particular, every positive operator of $\mathcal{B}'$ is essentially self-adjoint on $\mathcal{D}$. Let us show that such an S is affiliated to M'. One has, for $A \in M$, $x, y \in \mathcal{D}$,

$$(ASx, y) = (Ax, S^*y) = (Ax, S^+y)$$

since $S^* \subset S^+$. Since $\bar{S}$ is self-adjoint, $\bar{S} = S^+$, and by approximating elements of $\mathrm{Dom}(\bar{S})$ by suitable elements of $\mathcal{D}$, we get

$$(A\bar{S}x, y) = (Ax, \bar{S}y)$$

for $x \in \mathrm{Dom}(S)$, $y \in \mathrm{Dom}(\bar{S})$, $A \in M$. Hence, $x \in \mathrm{Dom}(\bar{S})$ imply $Ax \in \mathrm{Dom}(\bar{S}^+) = \mathrm{Dom}(\bar{S})$, and

$$(A\bar{S}x, y) = (\bar{S}^+Ax, y) = (\bar{S}Ax, y),$$

i.e., $A\bar{S} \subset \bar{S}A$, and $\bar{S}$ is affilied to M'. For any $S \geq Id$, $S \in (\mathcal{B}')^+$, one has $\bar{S}^2 = \overline{S^2}$ (lemma 9.1 of [1]). Coming back to the proof of lemma 8.3 of [1], for every $S \geq Id$, $S \in (\mathcal{B}')^+$, there exists n integer satisfying

$$\bar{S}^2 \leq \bar{A}_n^2 ,$$

hence

$$\bar{S}^2 \leq [\bar{A}_n]_Z^2 ,$$

where $[\bar{A}_n]_Z$ is a suitable operator self-adjoint affiliated to the center Z of M, with $[A_n]_Z \geq Id$ (Appendix [1]). One has $[A_n]_Z^{-1} \in Z \subset M'$, so that the operator $S' = S[\bar{A}_n]_Z^{-1}$ is bounded and satisfied $S = S'[A_n]_Z$; moreover, $S \in \mathcal{B}'$ and $Z \subset \mathcal{B}'$ give $S' \in \mathcal{B}'_{id}$. It is now obvious that elements of $\mathcal{B}'$ are linear combinations of positive elements S of $(\mathcal{B}')^+$, hence of strictly positive elements S of $(\mathcal{B}')^+$ (i.e., $S \geq \alpha > 0$ for some $\alpha > 0$). Finally, by theorem 1.1, $\mathcal{B}'$ is ultraweakly closed, relatively $D \hat{\otimes} D$, which achieves the proof of theorem 1.2.

We consistently use the obvious.

<u>Proposition 1.1</u>. 1°/ For a $*$-algebra A, operations $T \mapsto T^*$, $T \mapsto ST$, $T \mapsto TS$, with $T \in A, S \in A$, are weakly continuous and σ-weakly continuous.

2°/ For a $*$-algebra A with natural domain $\mathcal{D}$, and $x \in \mathcal{D}$, the map $T \in A \mapsto Tx$ is continuous from (A, λ) into the Hilbert space H completion of $\mathcal{D}$.

<u>Remark 1.1</u>. 1°/ At this point of the work, an important comment has to be made concerning a method often used in this paper. Let $A = \cup_{i\geq 0} A_{A_i}$ be a space with condition II and natural domain $\mathcal{D}$. Let i be fixed, and $A_i = \cup_{n\geq 0} A_{A_i^n}$ with natural domain $\mathcal{D}_i = \cap_{n\geq 0} \bar{A}_i^{-n}(H)$, where H is the Hilbert space completion of $\mathcal{D}$; clearly, A_i satisfies condition II, $\bar{A}_i^n(\mathcal{D}_i) = \mathcal{D}_i$ and $(A_i)_{id} = A_{id}$, as can be easily seen. The interest in the manipulation of A_i rests <u>on the quasi-normality of the Fréchet space</u> $\mathcal{D}_i$ (by corollary 3.1) whose central rôle is <u>a precise description of the ρ_i-convergence in terms of estimations</u> (note that (A_i, ρ_i) is finer than the topology of A_i induced by (A, ρ)). Many results concerning a general A with condition II are reduced to proofs concerning the A_i-spaces, and formulas $A = \cup_{i\geq 0} A_i$, $\mathcal{D} = \cap_{i\geq 0} D_i$, $A' = \cup_{i\geq 0} A'_i$ (by theorem 1.2), $D = \cap_{i\geq 0} D_i$ often lead to the desired result concerning A.

2°/ For a space A with natural domain $\mathcal{D}$, the commutant A' is contained in $B(\mathcal{D}, \mathcal{D})$ - the space of continuous sesquilinear forms on $\mathcal{D} \times \mathcal{D}$ - and is seen to be a $*$-algebra with domain $\mathcal{D}$ and a cofinal <u>central</u> sequence. It follows that there exists, for a given ultraweakly closed space A with condition II, a dissymmetry between $\mathcal{A}$ and $\mathcal{A}'$, since $\mathcal{A}$ has no algebraic structure, in general; more precisely, $\mathcal{A}'$ is determined in the algebraic commutation by a $*$-subalgebra $\mathcal{A}_0$ in $\mathcal{A}$ which suffices for commutation. In particular when the von Neumann algebra $M = \mathcal{A}_{id}$ appears to be a factor, $\mathcal{A}'$ is the usual commutant M' of M, as introduces in von Neumann theory. We shall show, in paragraph 10, that this commutant $\mathcal{A}'$ chosen in $B(\mathcal{D}, \mathcal{D})$ corresponds to the choice of the identity operator $Id \in M'$, and that other commutants for $\mathcal{A}$ may be introduced, each of them being associated to a suitable operator in M'. Thus, given two von Neumann algebra M and M' commuting each other in some Hilbert space, one may define many other commutants for M and each of them is an ultraweakly closed space with condition II containing M' and consisting of continuous sesquilinear forms on a new domain. In order to simplify these notes, we will defer the study of these commutants to paragraph 10, since they are not explicitely needed until this paragraph.

Proposition 1.2. Let $\mathcal{A}$ be an ultraweakly closed space with condition II, and $\mathcal{C} \subset \mathcal{A}$ be an arbitrary involutive subalgebra of $L(\mathcal{D})$, ultraweakly dense in $\mathcal{A}$. Given C in $L(\mathcal{D})$, one has $C \in \mathcal{A}'$ iff $(C\,Ax, y) = (Cx, A^*y)$ for all $A \in \mathcal{C}$ and $x, y \in \mathcal{D}$.

Proof. Let C in $L(\mathcal{D})$ with $(C\,Ax, y) = (Cx, A^*y)$, $A \in \mathcal{C}$, $x, y \in \mathcal{D}$. Since every $\beta \in \mathcal{A}$ is the ultraweakly limit of a suitable net $\beta_\alpha \in \mathcal{C}$, $\alpha \in I$ index set, we find, from formula $(Ax, y) = \lim_\alpha (A_\alpha x, y)$ for all $x, y \in \mathcal{D}$ with $A \in \mathcal{A} \cap L(\mathcal{D})$ and $A_\alpha \in \mathcal{C}$, $\alpha \in I$ (i.e., we take $\beta = A$) that $(ACx, y) = \lim_\alpha (A_\alpha Cx, y))$, $(Ax, C^*y) = \lim(A\,x, C^*y)$ due to $Cx \in \mathcal{D}$, $C^*y \in \mathcal{D}$, leading to $(ACx, y) = (Ax, C^*y)$ for all $A \in \mathcal{A} \cap L(\mathcal{D})$, hence $C \in \mathcal{A}'$. The converse is straightforward.

A linear operator T, densely defined, with domain Dom(T) is affiliated to a von Neumann algebra P (written $T \,\eta\, P$), iff, for any unitary operator U of P', one has $U\,T \subset T\,U$ (i.e., $U(\mathrm{Dom}(T)) \subset \mathrm{Dom}(T)$ and $UTx = TUx$ for all x in Dom(T)).

Take $\mathcal{B}_0$, $\mathcal{B}$, $\mathcal{B}'$, M, $\mathcal{D}$, H of theorem 1.2. If $t \in \mathcal{B}$ corresponds to some operator T such that $\mathrm{Dom}(T) \supset \mathcal{D}$ and $T(\mathcal{D}) \subset \mathcal{D}$ (via relation $t(x, y) = (Tx, y)$ for $x, y \in \mathcal{D}$), then it follows from

$$(T\,Bx, y) = (Tx, B^*y),$$

with $B \in M'$, that T with domain $\mathcal{D}$ is affiliated to M (since $B(\mathcal{D}) \subset \mathcal{D}$ and $B^*\mathcal{D} \subset \mathcal{D}$). Theorem 6.1 shows that such a situation is not always possible, namely, that there exists t in $\mathcal{B}$ which cannot correspond to operators T defined on $\mathcal{D}$. We often use

Lemma 1.4. Let T be closeable with dense domain Dom(T), and B a bounded operator such that $BT \subset TB$. Then $B\bar{T} \subset \bar{T}B$, where $\bar{T}$ is the closed operator associated to T.

Definition 1.2. Let $\mathcal{D}_1$ be some dense linear set of the Hilbert space H. A sesquilinear form β defined on $\mathcal{D}_1 \times \mathcal{D}_1$ is affiliated to M (written β η M) iff, for every unitary operator U of the commutant M' of M, one has $U\,\mathcal{D}_1 \subset \mathcal{D}_1$ and $\beta(Ux, Uy) = \beta(x, y)$ for all $x \in \mathcal{D}_1$, $y \in \mathcal{D}_1$.

Polarization equality shows that relation $\beta(Ux, Uy) = \beta(x, y)$ for all $x \in \mathcal{D}_1$, $y \in \mathcal{D}_1$ is equivalent to $\beta(Ux, Ux) = \beta(x, x)$ for all x in $\mathcal{D}_1$. It is not at all surprising that

Proposition 1.3. $\mathcal{B}$ is the set of β in $B(\mathcal{D}, \mathcal{D})$ which are affiliated to M.

Proof. Let $\beta = BA_i \times A_i$ in $\mathcal{B}$, with $B \in M$. For U unitary in M', one has $U(\mathcal{D}) = \mathcal{D}$ and $A_iU = UA_i$ on $\mathcal{D}$, therefore

$$\beta(Ux, Uy) = (BA_iUx, A_iUy) = (U^{-1}BUA_ix, A_iy) = \beta(x, y),$$

hence β η M. Conversely, let $\beta \in B(\mathcal{D}, \mathcal{D})$ with β η M. Then $\beta = BA_i \times A_i$ for some B in $L(H)$. From $A_i\mathcal{D} = \mathcal{D}$, we easily get $(BUx, Uy) = (Bx, y)$ for $x, y \in \mathcal{D}$, i.e., $B \in (M')' = M$ which is $\beta \in \mathcal{B}$.

Lemma 1.5. Let T be a closeable operator, T η M such that $\mathrm{Dom}(T) \supset \mathcal{D}$. Then, $T \in \mathcal{B}$.

Proof. We need to show that $T(Bx, y) = T(x, B^*y)$ for all $x, y \in \mathcal{D}$, and $B \in M'$. From $BT \subset TB$, we get, for $x \in \mathrm{Dom}(T)$ and $y \in H$,

$$(TBx, y) = (BTx, y),$$

which clearly implies for $x \in \mathcal{D} \subset \mathrm{Dom}(T)$, and $y \in \mathcal{D}$,

$$(TBx, y) = (Tx, B^*y),$$

hence the lemma is proved.

For sesquilinear forms already defined on $\mathcal{D} \times \mathcal{D}$, we will now mention

Proposition 1.4. Let f be a positive sesquilinear form on $\mathcal{D} \times \mathcal{D}$. Then

1°/ if there exists $g \in B(\mathcal{D}, \mathcal{D})$, such that $f \leq g$, one has $f \in B(\mathcal{D}, \mathcal{D})$.

2°/ if f is a closeable form, it must belong to $B(\mathcal{D}, \mathcal{D})$.

Proof. We will first prove 1°/. From the continuity of g on $\mathcal{D} \times \mathcal{D}$, one can find $M < +\infty$ and $i \in \mathbb{N}$, such that $|g(x, y)| \leq M \| A_i x\| \|A_i y\|$, hence

$$0 \leq f(x, x) \leq M(A_i x, A_i x)$$

The formula $g(u, v) = f(A_i^{-1}u, A_i^{-1}v)$ clearly defines a positive sesquilinear form, continuous on the pre-hilbert space $\mathcal{D} \times \mathcal{D}$, since $g(u, u) \leq M(u, u)$. Hence $g(u, v) = (Bu, v)$ for some $B \in L(H)$, which leads to $f = BA_i \times A_i$, i.e., $f \in B(\mathcal{D}, \mathcal{D})$.

For the second assertion, let f_1 be the closed form associated with f, with corresponding Hilbert space H_1 ([12], p. 313); by construction, we know that $H_1 \supset \mathcal{D}$. From the closed graph theorem, the canonical injection i from the Fréchet space $\mathcal{D}$ into the Hilbert space $(H_1, \| \|_1)$ is continuous. Indeed, relations $x_n \to 0$ in $\mathcal{D}$ and $i(x_n) \to y$ in H_1 imply $\| x_n \| \to 0$ in the Hilbert space H, and $f(x_n - x_m, x_n - x_m) \to 0$ when $n, m \to \infty$. Now, f_1 being closed, we get $f(x_n, x_n) \to 0$, hence $\|y\|_1 = 0$, that is, $y = 0$. Since H_1 is a Hilbert space, the Cauchy-Schwartz inequality is

$$|f(x, y)| \leq \|x\|_1 \ \|y\|_1$$

for $x, y \in H_1$, and continuity of i introduces an estimations of type $\|x\|_1 \leq M\|A_j x\|$ for some $j \in \mathbb{N}$ and $M < +\infty$ which gives, for $x, y \in \mathcal{D}$,

$$|f(x, y)| \leq M^2\|A_j x\| \ \|A_j y\|,$$

namely continuity of f on $\mathcal{D} \times \mathcal{D}$.

Lemma 1.6. Let $\theta = \theta^*$ be in $\mathcal{B}$. We assume that there exists $A \in \mathcal{B}^+$, $A \geq \mathrm{Id}$, and positive constants $\alpha, \beta > 0$, such that $\alpha A \leq \theta \leq \beta A$, and $A\mathcal{D} = \mathcal{D}$. Then, there exists an extension $\tilde{\theta}$ of θ to the product space $\mathrm{Dom}(\bar{A}^{\frac{1}{2}}) \times \mathrm{Dom}(\bar{A}^{\frac{1}{2}})$ and a self-adjoint operator $T \,\eta\, M$ such that $\tilde{\theta}(x, y) = (Tx, y)$ for $x \in \mathrm{Dom}(T) \subset \mathrm{Dom}(\bar{A}^{\frac{1}{2}})$ and $y \in \mathrm{Dom}(\bar{A}^{\frac{1}{2}})$. The square root self-adjoint operator $T^{\frac{1}{2}}$ is in $\mathcal{B}$, with $\mathrm{Dom}(T^{\frac{1}{2}}) = \mathrm{Dom}(\bar{A}^{\frac{1}{2}}) \supset \mathcal{D}$ and $\theta(x, y) = (T^{\frac{1}{2}}x, T^{\frac{1}{2}}y)$ for $x, y \in \mathcal{D}$.

Proof. It follows from the second commutant theorem, that $A^{-1} \in M$, $A^{\frac{1}{2}} \in \mathcal{B}$ with $A^{\frac{1}{2}}\mathcal{D} = \mathcal{D}$. We see, from relation $\alpha A \leq \theta \leq \beta A$, that there exists $B = B^*$ in M such that

$$\theta(x, y) = (B A^{\frac{1}{2}}x, A^{\frac{1}{2}}y)$$

for $x, y \in \mathcal{D}$, since $\mathcal{D} \subset \mathrm{Dom}(\bar{A}) \subset \mathrm{Dom}(\bar{A}^{\frac{1}{2}})$. The same relation shows that the form θ, with domain $D(\theta) = \mathcal{D}$ is closeable, with closure $\tilde{\theta}$ equal to

$$\tilde{\theta}(x, y) = (B \bar{A}^{\frac{1}{2}}x, \bar{A}^{\frac{1}{2}}y)$$

for $x, y \in D(\tilde{\theta}) = \mathrm{Dom}(\bar{A}^{\frac{1}{2}})$. One has, for $B \in A'$, $x, y \in \mathcal{D}$,

$$\theta(Bx, y) = \theta(x, B^*y)$$

and, for $B \in M'$, $BA \subset AB$, showing that

$$\tilde{\theta}(Bx, y) = \tilde{\theta}(x, By)$$

for $x, y \in \mathrm{Dom}(\bar{A}^{\frac{1}{2}})$ and $B = B^* \in M'$, since any element of $\mathrm{Dom}(\bar{A}^{\frac{1}{2}})$ may be approximated by suitable elements of $\mathcal{D}$. Now, it correspond to the form $\tilde{\theta}$, a self-adjoint operator $T \geq \alpha\mathrm{Id}$ such that $\mathrm{Dom}(T) \subset \mathrm{Dom}(\bar{A}^{\frac{1}{2}})$ and

$$\tilde{\theta}(x, y) = (Tx, y)$$

for $x \in \mathrm{Dom}(T)$, $y \in D(\tilde{\theta}) = \mathrm{Dom}(\bar{A}^{\frac{1}{2}})$. The square root $T^{\frac{1}{2}}$ satisfies $\tilde{\theta}(x, y) = (T^{\frac{1}{2}}x, T^{\frac{1}{2}}y)$ for $x \in D(\tilde{\theta})$, $y \in D(\tilde{\theta})$ and $\mathrm{Dom}(T^{\frac{1}{2}}) = D(\tilde{\theta})$ (for this we refer to [12]). Thus, $\mathrm{Dom}(T^{\frac{1}{2}}) = \mathrm{Dom}(A^{\frac{1}{2}})$ is stable under M', and we get, for $B \in M'$, $x, y \in \mathrm{Dom}(T^{\frac{1}{2}})$,

$$(T^{\frac{1}{2}}Bx, T^{\frac{1}{2}}y) = (T^{\frac{1}{2}}x, T^{\frac{1}{2}}By),$$

which implies that $T^{\frac{1}{2}}$, hence T, is affiliated to M, which is the lemma.

Proposition 1.5. 1°/ Let S and T be closeable operators, $S \eta M$, $T \eta M$ with $\mathrm{Dom}\, S \supset \mathcal{D}$ and $\mathrm{Dom}\, T \supset \mathcal{D}$. Then $(x, y) \in \mathcal{D} \in \mathcal{D} \longmapsto (Tx, Sy)$ belongs to $\mathcal{B}$.

2°/ Conversely, let $\beta = \beta^* \in \mathcal{B}$. One can find self-adjoint operators B_1, B_2, with $\mathrm{Dom}\, B_i \supset \mathcal{D}$, $B_i \,\eta\, M$ (for $i = 1, 2$), such that

$$\beta(x, y) = (B_1x, B_1y) - (B_2x, B_2y), \text{ for all } x, y \in \mathcal{D}$$

Proof. Take S, T of the first assertion. It has been already seen [11] that S (resp. T) is a continuous linear map from the Fréchet space $\mathcal{D}$ into the Hilbert space H, which leads to some estimations of type $\|Sx\| \le M_i\|A_i x\|$, $\|Tx\| \le M_j\|A_j x\|$ with $M_i, M_j < +\infty$.
From

$$|(Tx, Sy)| \le \|Tx\| \, \|Sy\| \le M_i M_j \| A_i x\| \, \|A_j x\|$$

follows continuity on $\mathcal{D} \times \mathcal{D}$. Now, take U a unitary operator in M'. One has U(Dom T) ⊂ Dom(T), and UTx = TUx for x in Dom(T) and similar properties of S. Then,

$$\beta(Ux, Uy) = (SUx, TUy) = (USx, UTy) = (Sx, Ty)$$

shows that β is affiliated to M, i.e., $\beta \in \mathcal{B}$
For the second assertion, we introduce $M < +\infty$ and $i \in \mathbb{N}$ such that $-M A_i^2 \le \beta \le M A_i^2$. Hence, $\beta + 2MA_i^2 = \theta$ satisfies lemma 1.6, which allows us to write
$\beta = -(2M)^{\frac{1}{2}} A_i \times (2M)^{\frac{1}{2}} A_i + T^{\frac{1}{2}} \times T^{\frac{1}{2}}$, proving the proposition.

A result, more or less contained in theorem 1.2, is

Proposition 1.6. Let $\mathcal{A} = \bigcup_{i\ge 0} \mathcal{A}_{A_i}$ be ultraweakly closed with condition II, and $\mathcal{A}_0 = \mathcal{A} \cap \mathcal{L}(\mathcal{D})$. For $\beta \in \mathcal{A}$ and $S \in \mathcal{A}_0$, one has $\beta(\,\cdot\,, S\,\cdot\,) \in \mathcal{A}$ and $\beta(S\,\cdot\,, \,\cdot\,) \in \mathcal{A}$.

Proof. Writing $\beta = B A_i \times A_i$ with $B \in \mathcal{A}_{id}$, we find that $\beta(\,\cdot\,, S\,\cdot\,) = B A_i \times A_i S \in \mathcal{B}(\mathcal{D}, \mathcal{D})$, and $\beta(S\,\cdot\,, \,\cdot\,) = B A_i S \times A_i \in \mathcal{B}(\mathcal{D}, \mathcal{D})$ by lemma 4.6 [1]. Taking $T = T^* \in (\mathcal{A}')_{id}$, one has, with $\gamma = \beta(\,\cdot\,, S\,\cdot\,)$, due to $T\mathcal{D} \subset \mathcal{D}$, for $x, y \in \mathcal{D}$,

$$\gamma(Tx, y) = B A_i T \times A_i S = B T A_i \times A_i S = TBA_i \times A_i S$$

$$\gamma(x, T^*y) = B A_i \times A_i ST = B A_i \times A_i TS =$$

$$= B A_i \times T A_i S = T B A_i \times A_i S,$$

since ST = TS on $\mathcal{D}$, $A_i T = T A_i$ on $\mathcal{D}$, hence $\gamma \in \mathcal{A}$. Similarly for $\gamma = \beta(S\,\cdot\,, \,\cdot\,)$.

CHAPTER 2: DENSITY THEOREMS

Lemma 2.1. Let $\mathcal{A}$ be a σ-weakly closed subspace of $B(\mathcal{D}, \mathcal{D})$, satisfying condition II, and E_α be a projector of the von Neumann algebra $\mathcal{A}_{id}$, such that $E_\alpha(H) \subset \mathcal{D}$. Then, for $\beta \in \mathcal{A}$, one has $\beta(E_\alpha \cdot , E_\alpha \cdot) \in \mathcal{A}_{id}$.

Proof. Let i be an integer and B an element of $\mathcal{A}_{id} = M$, such that $\beta = BA_i \times A_i$. One has $\beta(E_\alpha \cdot , E_\alpha \cdot) \in B(\mathcal{D}, \mathcal{D})$, from $E_\alpha(\mathcal{D}) \subset \mathcal{D}$ and, since A_iE_α is a bounded operator, it follows that $\beta(E_\alpha \cdot , E_\alpha \cdot) \in B(\mathcal{D}, \mathcal{D})_{id}$. For $A^* = A \in M'$, and x, y in $\mathcal{D}$, we get

$$\beta(E_\alpha Ax, E_\alpha y) = (BA_i E_\alpha Ax, A_iE_\alpha y)$$

and

$$(E_\alpha x, E_\alpha Ay) = (BA_iE_\alpha x, A_iE_\alpha Ay).$$

Since $\beta(Au, v) = \beta(u, Av)$, for $u, v \in \mathcal{D}$, and since $E_\alpha A = AE_\alpha$ on $\mathcal{D}$, by choosing $u = E_\alpha x$, $v = E_\alpha y$, we deduce from theorem 1.2 that $\beta(E_\alpha \cdot , E_\alpha \cdot) \in \mathcal{A}_{id}$.

Lemma 2.2. Let $\mathcal{A}$ be a subspace of $B(\mathcal{D}, \mathcal{D})$, stable under involution and satisfying the following properties (i) and (ii).

(i) there exists an increasing sequence of projectors (E_α) $\alpha \in \mathbb{N}$ in the Hilbert space H, such that $E_\alpha(H) \subset \mathcal{D}$, $(\cup_\alpha E_\alpha(H))^\perp = \{0\}$, and $(E_\alpha - Id)x$ goes to zero in the Fréchet space $\mathcal{D}$ (when $\alpha \to \infty$), for every $x \in \mathcal{D}$.

(ii) if $\beta \in \mathcal{A}$, then $\beta(E_\alpha \cdot , E_\alpha \cdot) \in \mathcal{A}$.

Then, $\mathcal{A}_{id}$ is weakly dense in $\mathcal{A}$. Moreover, if projectors E_α commute with a cofinal subset of $B(\mathcal{D}, \mathcal{D})$, $\mathcal{A}_{id}$ is σ-weakly dense in $\mathcal{A}$.

Proof. Let A_i be a cofinal sequence in $B^+(\mathcal{D}, \mathcal{D})$, and $\beta \in \mathcal{A}$. From [10], proposition 4, there is an operator $B \in \mathcal{L}(H)$, and an integer i such that

$\beta = BA_i \times A_i$. One has $\beta(E_\alpha \cdot , E_\alpha \cdot) \in B(\mathcal{D}, \mathcal{D})_{id}$, hence, by (ii), $\beta(E_\alpha \cdot , E_\alpha \cdot) \in A_{id}$.

Let us recall that $(E_\alpha x-x)$ tending to zero in $\mathcal{D}$ is equivalent to $\|A_i(E_\alpha x-x)\|$ tending to zero, for all i (when $\alpha \to \infty$). Taking x, y in $\mathcal{D}$, we get

$$\begin{aligned}|\beta(E_\alpha x, E_\alpha y) - \beta(x, y)| &= |(BA_iE_\alpha x, A_iE_\alpha y) - (BA_ix, A_iy)| \\ &\le |(BA_iE_\alpha x - BA_ix, A_iE_\alpha y)| + |(BA_ix, A_i(E_\alpha-Id)y)| \\ &\le \|B\| \|A_i(E_\alpha x-x)\| \|A_iE_\alpha y\| + \|B\| \|A_ix\| \|A_i(E_\alpha-1)y\|\end{aligned}$$

Since $A_iE_\alpha y$ tends to A_iy in the Hilbert space H, $\|A_iE_\alpha y\|$ is bounded, hence $\beta(E_\alpha \cdot , E_\alpha \cdot)$ tends weakly to β. When E_α commute with the A_i, the relation

$$|(BA_iE_\alpha x, A_iE_\alpha y)| \le \|B\| \|A_iE_\alpha x\|^2 \le \|B\|(A_i^2x, x)$$

shows that the set of $BA_iE_\alpha \times A_iE_\alpha$, for $\alpha \in \mathbb{N}$ is an equicontinuous set of $B(\mathcal{D}, \mathcal{D})$, therefore, by the Ascoli theorem, $\beta(E_\alpha \cdot , E_\alpha \cdot)$ tends ultraweakly to β.

Lemma 2.3. If there exists in $B^+(\mathcal{D}, \mathcal{D})$ a cofinal abelian sequence A_n (with $A_n \ge Id$, $A_n\mathcal{D} = \mathcal{D}$), then one can find a sequence E_n of projectors satisfying condition (i) of lemma 2 which belongs to the von Neumann algebra M generated by all A_n^{-1}.

Proof. Since M is abelian, M is finite, therefore $\mathcal{D}$ is essentially dense (see [17] or definition 3.2). Thus, there exist projectors E_n in M such that $E_n(H) \subset \mathcal{D}$, and $(\cup_n E_n(H))^\perp = \{0\}$. For x in $\mathcal{D}$, one clearly has $\|A_i(E_nx-x)\| = \|(E_n-Id)A_ix\|$ which tends to zero, since $E_n \to Id$ strongly.

One can now deduce

Proposition 2.1. Let A be a σ-weakly closed subspace of $B(\mathcal{D}, \mathcal{D})$, satisfying condition II. Then, the von Neumann algebra A_{id} is σ-weakly dense in A.

Proof. Let $\beta \in A$, and $B \in A_{id} = M$, $i \in \mathbb{N}$, such that $\beta = BA_i \times A_i$. Let us fix the integer i, and define

$\mathcal{D}_i = \bigcap_{n\geq 0} \text{Dom}(\bar{A}_i^n)$; $\mathcal{D}_i$ is the natural domain of the $*$-algebra generated by the powers A_i^n, for $n \in \mathbb{N}$, and since this cofinal subset is abelian, the relation $A_i\mathcal{D} = \mathcal{D}$ implies $\bar{A}_i(\mathcal{D}_i) = \mathcal{D}_i$. By construction, $\mathcal{D}$ is dense in the Fréchet space $\mathcal{D}_i$ and β has a unique extension to an element $\beta \in B(\mathcal{D}_i, \mathcal{D}_i)$; of course, $\bar{\beta} = B\bar{A}_i \times \bar{A}_i$ on $\mathcal{D}_i \times \mathcal{D}_i$. Put $\mathcal{B} = \bigcup_n M A_i^n \times A_i^n$. Then $\mathcal{B} \subset B(\mathcal{D}_i, \mathcal{D}_i)$, $\beta \in \mathcal{B}$ and, by theorem 1.1, $\mathcal{B}_{id} = M$. From lemma 2.1/2/3, β is the ultraweak limit, relative to $\mathcal{D}_i \hat{\otimes} \mathcal{D}_i$, of a sequence of elements of M, hence, is <u>a fortiori</u> the σ-weak limit, relative to $\mathcal{D} \hat{\otimes} \mathcal{D}$, of the above-mentioned sequence. This proves our proposition.

A simple extension of the Kaplansky density theorem is the following (see also theorem 11.2):

<u>Proposition 2.2</u>. Let $\mathcal{B}$ and $\mathcal{C}$ be subspaces of $B(\mathcal{D}, \mathcal{D})$, satisfying condition II with $\mathcal{B} \subset \mathcal{C}$, the cofinal sequence A_n being the same for $\mathcal{B}$ and $\mathcal{C}$. The following properties are equivalent:

1°/ $\mathcal{B}$ is σ-weakly dense in $\mathcal{C}$.

2°/ $\mathcal{B}_{id}$ is Hilbert σ-weakly dense in $\mathcal{C}_{id}$.

3°/ $\mathcal{B}_{A_n}$ is σ-weakly dense in $\mathcal{C}_{A_n}$, for some $n \geq 1$.

4°/ $[-A_n, A_n]_{\mathcal{B}}$ is σ-weakly dense in $[-A_n, A_n]_{\mathcal{C}}$, for some $n \geq 1$.

Let us recall that the Hilbert σ-weak topology in the usual σ-weak topology of von Neumann theory is the topology induced by the projective tensor product $H \hat{\otimes} H$ ('trace class operator'). Other formulations of proposition 2.2 can be obtained with weak topology in place of σ-weak topology.

<u>Proof</u>. Clearly, assertion 4°/ (resp. assertion 3°/) is satisfied iff it is satisfied for n = 1 ([1], lemma 5.4). Let us prove what 2°/ implies 3°/: it follows from Kaplansky density theorem that $[-Id, Id]_{\mathcal{B}}$, is Hilbert weakly dense in $[-Id, Id]_{\mathcal{C}}$. From the Ascoli theorem, topologies $\sigma(\mathcal{C}_{id}, \mathcal{D} \hat{\otimes} \mathcal{D})$ and $\sigma(\mathcal{C}_{id}, H \hat{\otimes} H)$ coïncide on $[-Id, Id]_{\mathcal{C}}$. Hence, $[-Id, Id]_{\mathcal{B}}$ is σ-weakly dense in $[-Id, Id]_{\mathcal{C}}$, thus $\mathcal{B}_{id}$ is σ-weakly dense in $\mathcal{C}_{id}$, which is 3°/. In the same way, the equivalence 2 $\Leftrightarrow$ 4 is an application of Kaplansky theorem, joined with Ascoli theorem. The implication 3°/ $\Rightarrow$ 1°/ is straightforward. Finally, if 1°/ is satisfied, one has $\mathcal{B}^{\sigma} = \mathcal{C}^{\sigma}$, hence $\mathcal{B}_{id}^{\sigma} = \mathcal{C}_{id}^{\sigma}$.

By theorem 1.1, $\mathcal{B}_{id}$ and $\mathcal{C}_{id}$ generated the same von Neumann algebra, thus the Kaplansky density theorem gives 1°/ ⇒ 2°/.

Lemma 2.4. Let A be an abelian $\star$-algebra satisfying condition I, and A^{σ} be its σ-weak closure. Then, the von Neumann algebra A^{σ}_{id} is ρ-dense and λ-dense in A.

Proof. Let T be in A, and A_i be a sequence cofinal in A^+. We can assume $T \geq Id$, since T is a linear combination of such elements. Clearly, the system (A_i) is cofinal in the commutant A', hence the natural domain D of A' coïncides with the domain $\mathcal{D}$ of A: it follows that A^{σ} is a $\star$-algebra with natural domain $\mathcal{D}$. As a consequence, $T\mathcal{D} = \mathcal{D}$, therefore T is essentially self-adjoint on $\mathcal{D}$. Let $\bar{T} = \int_1^{\infty} \lambda \, dp_{\lambda}$ be the spectral resolution of $\bar{T}$, and put $T_n = \int_1^n \lambda \, dp_{\lambda}$. One has $T_n \in A^{\sigma}_{id}$, $T_n \, \mathcal{D} \subset \mathcal{D}$, and the estimation

$$\|(T-T_n)x\| = \|TR_n T \, x| \leq \frac{1}{n}\|T^2 x\| ,$$

with $x \in \mathcal{D}$, and $R_n = \int_n^{\infty} \frac{1}{\lambda} \, dp_{\lambda}$; this proves our lemma.

Lemma 2.5. Let A be a $\star$-algebra satisfying condition I, and f, g be two positive linear forms on A, such that f = g on A_{id}. Then, f = g on A.

Proof. Let $(\pi_f, H_f, \mathcal{D}_f, \zeta_f)$ (resp., $(\pi_g, H_g, \mathcal{D}_g, \zeta_g)$) the G.N.S. representation of the $\star$-algebra A associated to f (resp. g). For T in A_{id}, one has

$$(\pi_f(T) \, \zeta_f, \zeta_f) = f(T) = g(T) = (\pi_g(T) \, \zeta_g, \zeta_g),$$

hence there exists a unitary operator U from H_f onto H_g such that $\pi_g(T)U = U\pi_f(T)$ for T in A_{id}. Since $\pi_f(A_i)\mathcal{D}_f = \mathcal{D}_f$, $\pi_g(A_i)\mathcal{D}_g = \mathcal{D}_g$, we get $U\mathcal{D}_f = \mathcal{D}_g$, $U\zeta_f = \zeta_g$. In the relation $\pi_g(A_i^{-1})Ux = U\pi_f(A_i^{-1})x$, replacing $x \in \mathcal{D}_f$ by $\pi_f(A_i)y$ with $y \in \mathcal{D}_f$, we get $\pi_g(A_i^{-1})U\pi_f(A_i)y = Uy$, hence $U\pi_f(A_i)y = \pi_g(A_i)Uy$ since $Uy \in \mathcal{D}_g$. Now let $\beta \in A$, with $\beta = B \, A_i \times A_i$ for some suitable B in A_{id} and $i \geq 0$. From $\pi_f(\beta) = \pi_f(B)\pi_f(A_i) \times \pi_f(A_i)$ and $f(\beta) = (\pi_f(B)\pi_f(A_i)\zeta_f, \pi_f(A_i)\zeta_f)$, and the corresponding formula for $\pi_g(\beta)$, $g(\beta)$, we deduce, for $\beta \in A$,

$$g(\beta) = (\pi_g(B)\pi_g(A_i)U\zeta_f, \pi_g(A_i)U\zeta_f)$$
$$= (\pi_g(B)U\pi_f(A_i)\zeta_f, U\pi_f(A_i)\zeta_f) = f(\beta).$$

Lemma 2.6. Let $\mathcal{A}$ be a $\star$-algebra satisfying condition I, and f be a continuous linear form on $\mathcal{A}$ such that $f(\mathcal{A}_{id}) = 0$. Then $f = 0$ on $\mathcal{A}$.

Proof. Since f^* also vanishes on $\mathcal{A}_{id}$, we can assume that f is hermitian. Since the positive cone $\mathcal{A}^+$ is normal, there exists f_1, f_2 positive linear forms on $\mathcal{A}$ such that $f = f_1 - f_2$ on $\mathcal{A}$. Clearly, $f_1 = f_2$ on $\mathcal{A}_{id}$, hence $f = 0$ by the preceding lemma.

Theorem 2.1. Let $\mathcal{A}$ be a $\star$-algebra satisfying condition I. Then, $\mathcal{A}_{id}$ is ρ-dense in $\mathcal{A}$.

Proof. Let $\mathcal{A}'$ be the dual space of $\mathcal{A}$ and $\langle\mathcal{A}, \mathcal{A}'\rangle$ be the corresponding duality. By lemma 2.6, the polar of $\mathcal{A}_{id}$ must be zero, hence $\mathcal{A}_{id}$ is dense in $\mathcal{A}$ for weak topology $\sigma(\mathcal{A}, \mathcal{A}')$. By the Minkowski theorem, the closure of $\mathcal{A}_{id}$ for ρ-topology coïncides with its weak closure, thus showing the theorem.

Proposition 2.3. 1°/ Let $\mathcal{A} = \bigcup_{i\geq 0} \mathcal{A}_{A_i} \subset B(\mathcal{D}, \mathcal{D})$, $\mathcal{A}$ stable under involution with $A_i \in \mathcal{A}$ for $i \geq 0$. For $S_1 \in L(\mathcal{D})$, $S_2 \in L(\mathcal{D})$, the map $\beta \in (\mathcal{A}, \rho) \to \beta(S_1 \cdot , S_2 \cdot) \in (B(\mathcal{D}, \mathcal{D}), \rho)$ is continuous. When $S_1(H) \subset \mathcal{D}$, $S_2(H) \subset \mathcal{D}$, this map is continuous from $(\mathcal{A}, \rho)$ into the Banach algebra $L(H)$, endowed with its Banach norm.

2°/ Let S_α be a net $(\alpha \in J)$ in $L(\mathcal{D})$ ρ-convergent to $S \in L(\mathcal{D})$ with $S_\alpha^* S_\alpha$ ρ-convergent to S^*S and contained in some $L(\mathcal{D})_{A_j}$. Then, for $B \in L(H)$, $B S_\alpha \times S_\alpha$ tends to $BS \times S$ in $(B(\mathcal{D}, \mathcal{D}), \rho)$. If $(A_iS_\alpha)^*(A_iS_\alpha)$ tends to $(A_iS)^* A_iS$ in $(L(\mathcal{D}), \rho)$ and is contained in some $L(\mathcal{D})_{A_k}$, then $\beta(S_\alpha \cdot , S_\alpha \cdot)$ is ρ-convergent to $\beta(S \cdot , S \cdot)$, for any $\beta \in \mathcal{A}_{A_i^2}$.

When J is countable, the sequence $S_n^*S_n$, being convergent to S^*S, defines a ρ-bounded set in $L(\mathcal{D})$, and hence is automatically contained in some $L(\mathcal{D})_{A_j}$.

Taking in 1°/ a projector E such that $E(H) \subset \mathcal{D}$, and a net β_α tending to β in (A, ρ), we find that the bounded operators $\beta_\alpha(E\cdot, E\cdot)$ tend to $\beta(E\cdot, E\cdot)$. A similar ≠ notion called $\star$-convergence has already been introduced and studied in [21].

Proof. 1°/ Due to polarization equality, we are reduced to the case where $S_1 = S_2 = S$. From $S\mathcal{D} \subset \mathcal{D}$, $S^*\mathcal{D} \subset \mathcal{D}$, and the closed graph theorem, we find that S is a continuous linear map from the Fréchet space $\mathcal{D}$ into itself. Writing $\beta = \beta^*$ as $\beta = BA_i \times A_i$ with $B \in L(H)$, we get $\beta(S\cdot, S\cdot) =$ $= BA_iS \times A_iS \in B(\mathcal{D}, \mathcal{D})$. From continuity of S, there exists $M < +\infty$ and $j \in \mathbb{N}$ such that $\|A_iSx\| \leq M\|A_jx\|$ for all $x \in \mathcal{D}$. Thus, $|\beta| \leq A_i^2$, equivalent to $\|B\| \leq 1$, leads to $|\beta(S\cdot, S\cdot)| \leq M^2A_j^2$, which is continuity of $\beta \to \beta(S\cdot, S\cdot)$. When $S_1(H) \subset \mathcal{D}$, $S_2(H) \subset \mathcal{D}$, we find $S(H) \subset \mathcal{D}$, thus, by the closed graph theorem, A_iS is a bounded operator in $L(H)$, and continuity with values in $L(H)$ follows from $|\beta(S\cdot, S\cdot)| \leq A_iS \times A_iS$.

2°/ For a net t_α ($\alpha \in J$, J index set) in $B(\mathcal{D}, \mathcal{D})$, such that $|t_\alpha| \leq \mu_\alpha$ for all $\alpha \in J$, where $\mu_\alpha \in B(\mathcal{D}, \mathcal{D})$ is a net ρ-convergent to zero, we find that t_α tends to 0 in $(B(\mathcal{D}, \mathcal{D}), \rho)$; indeed, $0 \leq t_\alpha + \mu_\alpha \leq 2\mu_\alpha$ and normality of the positive cone $B^+(\mathcal{D}, \mathcal{D})$ gives $(t_\alpha + \mu_\alpha) \to 0$, i.e., $t_\alpha \to 0$. From $(S_\alpha - S)^*(S_\alpha - S) = S_\alpha^*S_\alpha - S^*S_\alpha - S_\alpha^*S + S^*S$, we get that $(S_\alpha - S)^*(S_\alpha - S)$ tends to 0 in $(L(\mathcal{D}), \rho)$, or in $(B(\mathcal{D}, \mathcal{D}), \rho)$ due to $S(\mathcal{D}) \subset \mathcal{D}$, $S^*\mathcal{D} \subset \mathcal{D}$, to [1] corollary 1.b, and to 1°/ of this proposition. Since $S_\alpha^*\ S_\alpha$ is contained in some $L(\mathcal{D})_{A_j}$, there exists, for every $\alpha \in J$, a constant M_α such that $S_\alpha^*\ S_\alpha \leq M_\alpha A_j^2$, and we may assume that $S^*S \leq M\ A_j^2$ with $M < +\infty$; thus $\|S_\alpha x\| \leq M_\alpha^{\frac{1}{2}}\|A_jx\|$ for all $x \in \mathcal{D}$, and $\|Sx\| \leq M^{\frac{1}{2}}\|A_jx\|$. Clearly, $T_\alpha = S_\alpha\ A_j^{-1}$, $T = S\ A_j^{-1}$ are bounded operators of $L(\mathcal{D})$ and, from 1°/ of this proposition, $T_\alpha \to T$ and $T_\alpha^*T_\alpha \to T^*T$ in $(L(\mathcal{D}), \rho)$. As Id is an order-unit of A_{id}, we may take $B \geq 0$. Due to 1°/, we are reduced to show that $\gamma_\alpha = B\ T_\alpha \times T_\alpha$ is ρ-convergent to $\gamma = BT \times T$, since $BS_\alpha \times S_\alpha = \gamma_\alpha(A_j\cdot, A_j\cdot)$. In the equality

$$\gamma_\alpha - \gamma = B(T_\alpha - T) \times (T_\alpha - T) + BT \times T_\alpha + BT_\alpha \times T - 2BT \times T$$

the first term is ρ-convergent to zero, due to

$$0 \leq B(T_\alpha - T) \times (T_\alpha - T) \leq \|B\|(T_\alpha - T)^*(T_\alpha - T).$$

Let M be an equicontinuous subset consisting of positive linear forms on $B(\mathcal{D}, \mathcal{D})$. Each f in M induces a positive linear form f on the bounded involutive algebra $B(\mathcal{D}, \mathcal{D})_{id}$, thus

$$|f(T^*B(T_\alpha - T))|^2 \leq f(T^*BB^*T)f((T_\alpha - T)^*(T_\alpha - T))$$

From normality of $B^+(\mathcal{D}, \mathcal{D})$, topology ρ of $B(\mathcal{D}, \mathcal{D})$ is topology on $B(\mathcal{D}, \mathcal{D})$ of uniform convergence on positive equicontinuous sets, therefore, introducing $|\langle M, U\rangle| = \sup\{|g(U)| \ ; \ g \in M\}$ for $U \in L(H)$, we get that the net $|\langle M, T^*B(T_\alpha - T)\rangle|$ tends to zero, showing that $T^*B(T_\alpha - T) = B(T_\alpha - T) \times T$ is ρ-convergent to zero. In the same way, $BT \times T_\alpha \to BT \times T$, leading to $(\gamma_\alpha - \gamma) \to 0$ in $(B(\mathcal{D}, \mathcal{D}), \rho)$. Now let $\beta \in A_{A_i^2}$: then, $\beta = B\,A_i \times A_i$ for some $B \in L(H)$. Clearly, $\beta(S_\alpha \cdot\ , S_\alpha \cdot\ ) = B\,A_i\,S_\alpha \times A_i\,S_\alpha$, $\beta(S \cdot\ , S \cdot\ ) = B\,A_iS \times A_iS$. From our assumptions, $A_iS_\alpha \to A_iS$ in $(L(\mathcal{D}), \rho)$ as well as $(A_iS_\alpha)^*(A_iS_\alpha) \to (A_iS)^*(A_iS)$. Due to $(A_iS_\alpha)^*(A_iS_\alpha) \in L(\mathcal{D})_{A_k}$, we find that the net $BA_i\,S_\alpha \times A_iS$ tends to $BA_iS \times A_iS$ in $(B(\mathcal{D}, \mathcal{D}), \rho)$, thus proving the proposition.

Theorem 2.2. Let A be a space satisfying condition II, with natural domain $\mathcal{D}$. Then, A_{id} is ρ-dense in A.

For example, for an ultraweakly closed space A satisfying condition II, the von Neumann algebra A_{id} is ρ-dense in A. Taking $A = B(\mathcal{D}, \mathcal{D})$, we find that the Banach algebra $L(H)$ is ρ-dense in $B(\mathcal{D}, \mathcal{D})$.

Proof. Let $\beta \in A$, and $B \in A_{id}$, $j \in \mathbb{N}$ such that $\beta = BA_j \times A_j$. The integer j being fixed, we put $\mathcal{D}_j = \bigcap_{n\geq 0} \mathrm{Dom}(\bar{A}_j^n) = \bigcap_{n\geq 0} A_j^{-n}(H)$ and $\mathcal{B} = \bigcup_{n\geq 0} A_{id}\,A_j^n \times A_j^n$; clearly, $A_{id} \subset \mathcal{B}_{id}$ and $\mathcal{B} \subset A$ implies that $\mathcal{B}_{id} \subset A_{id}$ so that $\mathcal{B}$ satisfies condition II and has natural domain $\mathcal{D}_j$. The natural embedding from $(\mathcal{B}, \rho)$ into (A, ρ) being continuous, it is sufficient to show the theorem for $\mathcal{B}$.

Let $\mathcal{C}$ be the abelian $*$-algebra with natural domain $\mathcal{D}_j$ generated by A_j, A_j^{-1}. Then, $\mathcal{C}_{id} \subset \mathcal{B}_{id}$, $\mathcal{C}$ satisfies condition I and $(\mathcal{B}, \rho)$ induces $(\mathcal{C}, \rho)$ on $\mathcal{C}$. By lemma 2.4, $\mathcal{C}_{id}$ is ρ-dense in $\mathcal{C}$, therefore there exists a net S_α in $\mathcal{B}_{id}$ such that S_α tends to A_j for ρ-topology. Noting that topologies λ and ρ are equal on $\mathcal{C}$, we get that $S_\alpha^* S_\alpha$ tends to A_j^2 in $(\mathcal{C}, \rho)$. From proposition 2.2. 2°/, $BS_\alpha \times S_\alpha$ tends to $B\, A_j \times A_j$, due to $S_\alpha^* S_\alpha \in \mathcal{B}_{id}$, and clearly $BS_\alpha \times S_\alpha \in \mathcal{B}_{id}$.

Remark 2.1. The density of $A_{id} \cap L(\mathcal{D})$ or of $A \cap L(\mathcal{D})$ is a more powerful property and the reader is referred to lemma 7.8.

Lemma 2.7. In an ultraweakly closed space $A = \bigcup_{j \geq 0} A_{A_j}$ with condition II (and $M = A_{id}$) the following properties are equivalent.

1°/ For any $\beta \geq 0$ in A, $\beta \neq 0$, there exists $B \neq 0$, $B \in M$ such that $0 \leq B \leq \beta$.

2°/ For any $\beta \geq 0$ in A, $\beta \neq 0$, $[0, \beta]_A \cap M$ is ultraweakly dense in $[0, \beta]_A$.

3°/ for any $\beta \geq 0$, there exists an increasing net F in M^+ with β as least upper bound.

Such properties are satisfied when A_i is a central sequence.

When A_i is not central, properties mentionned are realized for those β which commute with a given A_j.

Proof. Equivalences of 1°/, 2°/, 3°/ follow from Zorn's lemma and the ultraweak compacity of intervals $[-A_i, A_i]$, $i \geq 0$. Let us assume now that A_i is central; any $0 \leq \beta = BA_i \times A_i$ in A is the ultraweak limit, as $\alpha \to \infty$, relative to $\mathcal{D}_i \hat{\otimes} \mathcal{D}_i$ of $\beta_\alpha = \beta(E_\alpha \cdot\, , E_\alpha \cdot\,)$, as introduced in lemma 2.2, and $0 \leq \beta_\alpha \leq \beta$ on $\mathcal{D}_i$ where $E_\alpha(H)_\alpha \subset \bigcap_{n \geq 0} A_i^{-n}(H) = \mathcal{D}_i$. The natural domain of $\mathcal{B}_i = \bigcup_{n \geq 0} MA_i^n \times A_i^n$ is $\mathcal{D}_i$ and the imbedding $j_i : \mathcal{B}_i \to A$ is ρ-continuous and normal, hence ultraweakly continuous from $\sigma(\mathcal{B}_i, \mathcal{D}_i \hat{\otimes} \mathcal{D}_i)$ into $\sigma(A, \mathcal{D} \hat{\otimes} \mathcal{D})$, implying that β_α tends to β for ultraweak topology of A.

Many consequences of density may be drawn. For example, let A of theorem 2.2 and $\mathcal{B}$ be a subspace of some $B(\mathcal{D}_\pi, \mathcal{D}_\pi)$ where $\mathcal{D}_\pi$ is a given dense domain - with a natural Fréchet topology -

in a second Hilbert space. For a continuous linear map from $(\mathcal{A}, \rho)$ into $(\mathcal{B}, \rho)$, we see that $\pi \geq 0$ on $\mathcal{A}_{id}$ is equivalent to $\pi \geq 0$ on $\mathcal{A}$.

Let $\mathcal{A}$ be an ultraweakly closed space with condition II, $\mathcal{A} \subset B(\mathcal{D}, \mathcal{D})$ with commutant $\mathcal{A}'$ in $B(\mathcal{D}, \mathcal{D})$. One can checked that $\mathcal{A}'_{id}$ is ρ-dense and λ-dense in $\mathcal{A}'$, hence σ-weakly dense for all possible σ-topologies on $\mathcal{A}'$ (i.e., relatively to $\mathcal{D} \hat{\otimes} \mathcal{D}$, $D \hat{\otimes} D$, $D \hat{\otimes} H$, and $\mathcal{D} \hat{\otimes} H$ where D is the natural domain of $\mathcal{A}'$; note that $\mathcal{D} \subset \cup_{A \in \mathcal{A}_{id}} A(\mathcal{D}) \subset D$).

We will need

Proposition 2.4. Let $\mathcal{A}$ be a space with condition II and cofinal sequence A_i, and f be a continuous linear form on $\mathcal{A}$. We assume that there exists f_1, f_2 positive linear forms on $\mathcal{A}$ and a particular value of the integer i such that $f = f_1 - f_2$ on $\mathcal{A}_{id}$(*) with

$$\|f\|_{A_i^2} = \|f_1\|_{A_i^2} + \|f_2\|_{A_i^2} \quad (**).$$

Then $f = f_1 - f_2$ on whole $\mathcal{A}$ and (f_1, f_2) is the unique couple of positive linear forms on $\mathcal{A}$ satisfying (*) and (**).

Recall that, for $g \in \mathcal{A}^\rho$, $\|g\|_{A_j}$ denotes the quantity

$$\|g\|_{A_j} = \sup_{T \in [-A_j, A_j]} |g(T)|$$

and that the sequence of semi-norms $g \in \mathcal{A} \longmapsto \|g\|_{A_n}$ or $\|g\|_{A_n^2}$ defines the topology of the strong dual $\mathcal{A}^\rho$ of $\mathcal{A}$.

Proof. Linear forms f and $f_1 - f_2$ being continuous on $\mathcal{A}$ must coïncide on $\mathcal{A}$ by density results. Maps $\beta \longmapsto A_i \beta A_i$ and $\beta \longmapsto A_i^{-1} \beta A_i^{-1}$ are ρ-homeomorphisms and, by lemma 1.1, exactly exchange $\mathcal{A}_{id}$ and $\mathcal{A}_{A_i^2}$, thus we are reduced to the case where $A_i = \mathrm{Id}$. The restriction of f to $\mathcal{A}_{id}$ being continuous for $\| \ \|_{id}$, formulas (*) and (**) hold on the $\mathbb{C}^*$-algebra generated by the involutive algebra $\mathcal{A}_{id}$ for which unicity is well-known.

CHAPTER 3: TECHNICAL PROPERTIES OF THE DOMAIN

Let E be a locally convex topological vector space; E is said to be quasi-normable if, for every convex neighborhood U of zero, there exists a neighborhood V of zero such that, for every $\varepsilon > 0$, there exists a bounded set M_ε in E satisfying $V \subset \varepsilon U + M_\varepsilon$. In particular, it is easily seen that every subspace $\mathcal{A} \subset B(\mathcal{D}, \mathcal{D})$ with $\mathcal{A} = \mathcal{A}^*$, endowed with topology ρ, is quasi-normable.

The author does not know in general if the domain $\mathcal{D}$ of any $*$-algebra $\mathcal{A}$ is a quasi-normable Fréchet space: as seen in [10], this property plays a fundamental role for analytical developments. The beginning of this paragraph is a partial answer to this central problem.

A fundamental system of zero neighborhoods of the Fréchet space $\mathcal{D}$ consists of sets U of type $U = \{x \in \mathcal{D} |\ \|Ax\| \leq 1\}$, with A moving in $\mathcal{A}$. Thus, the relation $V \subset \varepsilon U + M_\varepsilon$ remains to show that, for any $A \in \mathcal{A}$, there exists $B \in \mathcal{A}$ such that, for every $\varepsilon > 0$, there exists a bounded set β in $\mathcal{D}$ with the following property: any $x \in \mathcal{D}$ with $\|Bx\| \leq 1$ can be written $x = x_1 + x_2$, with $x_1, x_2 \in \mathcal{D}$, $\|Ax_1\| \leq \varepsilon$, and $x_2 \in \beta$.

Let us consider a $*$-algebra $\mathcal{D}$ satisfying condition I, with natural domain $\mathcal{D}$. It follows from the classical isomorphism theorem between Fréchet spaces that any element C of $\mathcal{A}$, such that $C\mathcal{D} = \mathcal{D}$, induces a topological isomorphism from the Fréchet space $\mathcal{D}$ into itself, implying that the map $C: \mathcal{D} \to \mathcal{D}$ exchanges bounded sets of $\mathcal{D}$. We denote by Ω the trace on $\mathcal{D}$ of the unit ball H_1 of the Hilbert space H, and by A, B, ... elements of a suitable cofinal subset of $\mathcal{A}^+$. Our condition then becomes: for any $A \in \mathcal{A}$, there exists $B \in \mathcal{A}$ such that, for any $\varepsilon > 0$, and $y \in \Omega = H_1 \cap \mathcal{D}$, there exists a decomposition of the vector $AB^{-1}(y)$ in the form

$$A\,B^{-1}y = y_1 + y_2$$

with $y_1 \in \mathcal{D}$, $\|y_1\| \leq \varepsilon$, and y_2 moving in a bounded set of $\mathcal{D}$ (independent of y). The notation B^{-1} is unambiguous due to $B\mathcal{D} = \mathcal{D}$. Moreover, it is possible to replace AB^{-1} by its absolute value $|AB^{-1}|$; indeed, putting $T = AB^{-1}$, we see that $T^* \supset B^{-1}A$, hence $|T|(\mathcal{D}) = \mathcal{D}$, $|T^*|(\mathcal{D}) = \mathcal{D}$, therefore the unitary operator U appearing in the polar decomposition of T satisfies $U(\mathcal{D}) = \mathcal{D}$, so that U is a topological isomorphism from $\mathcal{D}$ into itself, thus proving our assertion.

Lemma 3.1. Let M be a linear subset of $\mathcal{D}$, and $T \in A$, with $T \geq Id$, $T\mathcal{D} \subset \mathcal{D}$. If M is complete for topology defined by the norm $x \mapsto \|Tx\|$, then the unit ball of M for this topology is a bounded subset of the Fréchet space $\mathcal{D}$.

Indeed, it follows from the closed graph theorem and the fact that T is closeable on $\mathcal{D}$, that the canonical injection for the Banach space M into the Fréchet space $\mathcal{D}$ is continuous, which gives the lemma.
A variant on lemma 3.1 is the easy

Lemma 3.2. Let M be a closed linear subspace of the Hilbert space H, with $M \subset \mathcal{D}$. Then, M is closed in the Fréchet space $\mathcal{D}$ and both topologies on M (i.e., induced by H and $\mathcal{D}$) coïncide.

Certain bounded subsets of $\mathcal{D}$ can be obtained by

Proposition 3.1. Let A be a $*$-algebra on which λ coïncides with ρ, and B be a bounded subset of the Fréchet space $\mathcal{D}$. For any bounded subset S of (A, λ), the set $\cup_{T \in S} T(B)$ is bounded in $\mathcal{D}$.

Proof. We need to show that the subset involved is simply bounded on $\mathcal{D}'$. If $f \in \mathcal{D}'$ there exists $j \in \mathbb{N}$, $y \in H$ such that $f(u) = (A_j u, y)$ for all $u \in \mathcal{D}$. Therefore, for $T \in S$,

$$<T(B), f> = <A_j T(B), y>$$

and the set of $A_j T$ being bounded in (A, λ) by [11], one can find k such that $\|A_j Tx\| \leq \|A_k x\|$ for all $x \in \mathcal{D}$, which is our proposition.

Our main result is

Proposition 3.2. Let $\mathcal{D}$ be the natural domain of a $*$-algebra satisfying condition I. If there exists in the Hilbert space H a symmetric operator (not necessarily bounded) T such that $T(\mathcal{D}) = \mathcal{D}$, and bounded operators p_n, $n \in \mathbb{N}$ such that $p_n(H) \subset \mathcal{D}$ and $\|T^{-1}(1-p_n)\| \to 0$ when $n \to +\infty$, then the Fréchet space $\mathcal{D}$ is quasinormable.

Proof. It is immediately seen that, for an operator S such that $S(\mathcal{D}) \subset \mathcal{D}$, $s^*(\mathcal{D}) \subset \mathcal{D}$ the semi-norm $x \mapsto \|Sx\|$ is continuous on $\mathcal{D}$. Let $A \in A$, and $\varepsilon > 0$; taking $B = TA$, one clearly has $B\mathcal{D} = \mathcal{D}$ and $AB^{-1} = T^{-1}$ on $\mathcal{D}$. Taking an integer n such that

$\|T^{-1}(1-p_n)\| \leq \varepsilon$, we get, for $y \in \Omega$,

$$AB^{-1}y = T^{-1}y = Tp_n y + T^{-1}(1-p_n)y,$$

With $y_1 = T^{-1}(1-p_n)y$ and $y_2 = T^{-1}\ p_n y$, it follows that, for y in Ω,

$$\|y_1\| \leq \|T^{-1}(1-p_n)\|\ \|y\| \leq \varepsilon.$$

By the closed graph theorem, $T^{-1}p_n$ is linear continuous from the Hilbert space H into the Fréchet space $\mathcal{D}$, hence $T^{-1}\ p_n\ \Omega$ is bounded in $\mathcal{D}$, thus proving our proposition.

Corollary 3.1. Let $\mathcal{D}$ be a dense linear subset of some Hilbert space H, and A be a linear operator defined on $\mathcal{D}$, such that $A \geq \mathrm{Id}$, $A\ \mathcal{D} \subset \mathcal{D}$, and such that all powers A^n are essentially self-adjoint on $\mathcal{D}$. Put $\hat{\mathcal{D}} = \cap_{n\geq 0} \mathrm{Dom}(\bar{A}^n)$. Then $\bar{A}^n\ \hat{\mathcal{D}} = \hat{\mathcal{D}}$ for $n \in \mathbb{N}$, and $\hat{\mathcal{D}}$ is quasi-normable for topology defined by the seminorms $x \to \|\bar{A}^n x\|$.

If $\mathcal{B}$ is some $\star$-algebra with domain $\mathcal{D}$ admitting the sequence A^n as a cofinal subset, then the commutant $\mathcal{B}'$ with natural domain D(D $\supset \mathcal{D}$ by [1]) has a similar description; namely, the operator $B = A^Z$ introduced in [1] and its powers B^n are cofinal in $\mathcal{B}'$, with $B^n D = D$. Of course, whenever there exists in some $\star$-algebra $\mathcal{A}$ a finite number A_i, $1 \leq i \leq p$ of elements with $A_i\mathcal{D} = \mathcal{D}$ and $A_iA_j = A_jA_i$ for $1 \leq i, j \leq p$, such that all possible powers $A_1^{i_1} \ldots A_p^{i_p}$, with $(i_1,\ldots,i_p) \in \mathbb{N}^p$, contain a cofinal subset of $\mathcal{A}^+$, we are brought back to the case just considered, because the powers A^n of the operator $A = A_1A_2 \ldots A_p$ are clearly cofinal in $\mathcal{A}^+$.

Such considerations appear when $\mathcal{D}$ is the set of differentiable vectors obtained from a continuous unitary representation of a real Lie group, with A equal to the Laplacian; hence $\mathcal{D}$ is quasi-normable (see example 2°/).

Proof. The fact that $\bar{A}^n\ \hat{\mathcal{D}} = \hat{\mathcal{D}}$ follows from self-adjointness. Let $\bar{A}^{-1} = \int_0^1 \lambda\ dp_\lambda$ be the spectral decomposition of the closure $\bar{A}^{-1}$ and $p_n = \int_{1/n}^1 dp_\lambda$, for $n \in \mathbb{N}$. For $k \geq 0$, one has $\bar{A}^{-k} \geq 1/n^k$ on the subspace $p_n(H)$, hence $p_n(H) \subset \mathrm{Dom}(\bar{A}^k)$, i.e., $p_n(H) \subset \mathcal{D}$. Now $\|T^{-1}(1-p_n)\| \leq 1/n$, so that proposition 3.2 can be applied.

We now develop other properties of the domain, useful for analysis.

Proposition 3.3. Let $\mathcal{D}$ be the natural domain of a (general) $*$-algebra $\mathcal{A} = \cup_{i\geq 0} \mathcal{A}_{A_i}$ and L be a continuous linear form on $\mathcal{D}$. Then, there exists an integer i, and $\zeta \in \mathrm{Dom}(\bar{A}_i)$ such that

$$L(\varphi) = (\bar{A}_i \varphi, \bar{A}_i\zeta) = (\bar{A}_i^2 \varphi, \zeta), \quad \text{for all } \varphi \in \mathcal{D}.$$

It is easily seen that L moves in an equicontinuous subset of $\mathcal{D}'$ iff ζ moves in a subset of $\mathrm{Dom}(\bar{A}_i)$ bounded for the norm $\zeta \in \mathrm{Dom}(\bar{A}_i) \longmapsto \|\bar{A}_i\zeta\|$.

Definition 3.1. A continuous linear form L on $\mathcal{D}$ is of order i iff there exists $\zeta \in \mathrm{Dom}\, \bar{A}_i$ such that $L(x) = (A_i x, \bar{A}_i\zeta)$ for all $x \in \mathcal{D}$.

Clearly, L is of order i iff $\|Lx\| \leq C^{te}\|A_i x\|$ for all $x \in \mathcal{D}$. The map $\zeta \in \mathrm{Dom}\, \bar{A}_i \longmapsto L$ exchanges bijectively $\mathrm{Dom}\, \bar{A}_i$ and the set of continuous linear forms of order i.
Note that $L \in \mathcal{D}'$ iff there exists $i \geq 0$ such that $L(A_i^{-1} \cdot\,) \in H'$. For a Schwartz space $\mathcal{D}$ [10] we see, from proposition 3.3 and diagonalization of a suitable A_j^{-1}, that there exists, for a given $L \in \mathcal{D}'$ (resp., L moving in an equicontinuous set in $\mathcal{D}'$) an orthonormal basis e_n in H, a sequence (λ_n) of positive real $\to 0$ and a vector $w \in H$ (resp., w moving in a bounded set of H) such that $L(v) = \Sigma_n\lambda_n(v_n, w_n)$ for all $v \in \mathcal{D}$, where $v = (v_n)$, $w = (w_n)$. Conversely, the equicontinuous principle shows that every such expansion, absolutely converging for all $v \in \mathcal{D}$, defines an element of $\mathcal{D}'$. For a Schwartz space $\mathcal{D}$ of the form $\mathcal{D} = \cap_{n\geq 0} \Delta^{-n}(H)$ (example 2°/), the same basis (e_n) may be chosen for all $L \in \mathcal{D}'$; note that $e_n \in \mathcal{D}$ for $\lambda_n \neq 0$.

Proof. There exists an integer i and $M < +\infty$ such that $\|L\varphi\| \leq M\|A_i\varphi\|$ with $\|A_i\varphi\| \geq \|\varphi\|$ for all φ in $\mathcal{D}$. We will assume that $M = 1$. Clearly, $\mathrm{Dom}(\bar{A}_i)$ is a Hilbert space for the scalar product $(x, y) \in \mathrm{Dom}(\bar{A}_i) \times \mathrm{Dom}(\bar{A}_i) \longmapsto (\bar{A}_i x, \bar{A}_i y)$. From the preceding formula, L has a continuous extension $\bar{L}$ to $\mathrm{Dom}(\bar{A}_i)$ and $\|\bar{L}\varphi\| \leq \|\bar{A}_i\varphi\|$ for all $\varphi \in \mathrm{Dom}(\bar{A}_i)$. The linear set $\bar{A}_i(\mathrm{Dom}(\bar{A}_i))$ is a Hilbert space K_i, for the scalar product of H, so that our formula becomes

$$|\bar{L}_0 \bar{A}_i^{-1} (v)| \leq \|v\|$$

for $v \in K_i$, and the Riesz Fischer theorem gives some vector $u \in K_i$ such that $L(\varphi) = (A_i\varphi, u)$. Since $u = \bar{A}_i\zeta$ for some ζ in $\mathrm{Dom}(\bar{A}_i)$, we prove our proposition.

Remark 3.1. This remark claims no originality and considers integral formulas for sesquilinear forms $\beta = \beta^* \in B(\mathcal{D}, \mathcal{D})$. For a nuclear space $\mathcal{D}$ one may use a general kernel theorem (attributed to Schwartz-Grothendieck), expanding β as an infinite serie $\beta = \Sigma_i \lambda_i x_i' \otimes y_i'$ with $\Sigma_i \lambda_i < +\infty$ $(x_i')_{i\geq 0}$ (resp. $(y_i')_{i\geq 0}$) moving in equicontinuous sets in $\mathcal{D}'$, which may be proved in our context as follows. First, choose $j \geq 0$ sufficiently large such that A_j^{-1} is nuclear in H [10], proposition 2) and such that $\beta \in B(\mathcal{D}, \mathcal{D})_{A_j^2}$. Thus, we get $B = B^* \in L(H)$ such that $\beta = BA_j \times A_j$ and B may be assumed to be nuclear in H, due to $\beta = (A_j^{-1}BA_j^{-1})A_j^2 \times A_j^2$. Choosing φ_i $i \geq 0$ an orthonormal basis H consisting of eigenvectors for B (i.e., $B\varphi_i = \lambda_i\varphi_i$ for all i, and $\Sigma_i\lambda_i < +\infty$) we get, for $u, v \in \mathcal{D}$,

$$\beta(u, v) = \sum_{i=1}^{\infty} \lambda_i (A_j u, \varphi_i)\overline{(A_j v, \varphi_i)}$$

and the sequence $x_i' \equiv (A_j \cdot , \varphi_i)$ $i \geq 0$ is clearly equicontinuous in H since $\|\varphi_i\| \leq 1$ for all $i \geq 0$ (proposition 3.3). We note that $\beta = \beta^*$ implies $x_i' = y_i'$, and the case $\beta \geq 0$ gives $\lambda_i \geq 0$ for all i.
Nuclearity of $\mathcal{D}$ ensures that $\Sigma_i|\lambda_i| < +\infty$ and we see that the expansion of β considered holds even if $\mathcal{D}$ is simply a Schwartz space; here, A_j^{-1} may be choosen compact and $\Sigma_i|\lambda_i|$ may diverge; the formula so obtained is more closed to finite dimensional decomposition of quadratic forms.

The expansion introduced in the Kernel theorem is obviously not unique, and may be written as

$$\beta = \int_{\lambda \in \mathrm{sp}(B)} \lambda \, d(p_\lambda A_j \times A_j),$$

where p_λ are spectral projectors of B, noting that

$$(p_\lambda A_j \times A_j)(u, v) = (A_j u, \varphi_\lambda)\overline{(A_j v, \varphi_\lambda)}.$$

When $\mathcal{D}$ is not nuclear, this presentation of β as an integral still holds. Indeed, representation $\beta = BA_j \times A_j$ and von Neumann's spectral theorem $B^* = B = \int \lambda \, dp_\lambda$ leads to the same formula $\beta = \int \lambda \, d(p_\lambda A_j \times A_j)$. This may be viewed as an integral decomposition over the set of extremal rays of the positive cone $B^+(\mathcal{D}, \mathcal{D})$ in $B(\mathcal{D}, \mathcal{D})$; indeed, an extremal ray λ of $B^+(\mathcal{D}, \mathcal{D})$ must lie in some $B(\mathcal{D}, \mathcal{D})_{A_j^2}$ for suitable $j \geq 0$, and remains extremal in the positive cone of $B(\mathcal{D}, \mathcal{D})_{A_j^2}$, thus has the form $\lambda = pA_j \times A_j$, since $B \in L(H) \longmapsto BA_j \times A_j \in B(\mathcal{D}, \mathcal{D})_{A_j^2}$ is an order isometry and, therefore, preserves extremal rays. Conversely each $pA_j \times A_j$ is clearly extraml in $B^+(\mathcal{D}, \mathcal{D})$. An equivalent, but different, version is the following. Let H_{A_j} be the Hilbert space Dom $\bar{A}_j$ with scalar product $(x, y)_{A_j} \equiv (\bar{A}_j x, \bar{A}_j y)$ for $x, y \in H_{A_j}$. From $|\beta(x, y)| \leq$ $\leq C^{te} \|A_j x\| \, \|A_j y\|$, we see that $\beta = BA_j \times A_j$ corresponds to a bounded operator S_β from H_{A_j} into itself, i.e.,

$$\beta(x, y) = (S_\beta x, y)_{A_j} = (\bar{A}_j S_\beta x, \bar{A}_j y).$$

The map $S \in L(H_{A_j}) \longmapsto B = \bar{A}_j S A_j^{-1} \in L(H)$ is a bijective isometry, i.e., $S = A_j^{-1} B \bar{A}_j$ and spectral theorem $S_\beta = \int \lambda \, dG_\lambda$, with G_λ closed spaces in H_{A_j}, leads to the same formula $\beta(x, y) = \int \lambda \, d(\bar{A}_j G_\lambda \times \bar{A}_j y)$. Thus, β may be viewed as an integral involving closed sets of the Hilbert space Dom $\bar{A}_i$. Representations of β involving two parameters λ, μ, or involving an integral with distributions evaluations, are probably valuable.

For an element $A \in A$, such that $\|Ax\| \geq \|x\|$ for all $x \in \mathcal{D}$, we denote by H_A the Hilbert space $\mathrm{Dom}(\bar{A})$, with scalar product $(x, y)_A = (\bar{A}x, \bar{A}y)$, for $x, y \in \mathrm{Dom}(\bar{A})$. For A, B in A with $\|Bx\| \geq \|Ax\| \geq \|x\|$ for $x \in \mathcal{D}$, one has $H_B \subset H_A$ and the first space embeds continuously in the second.

Proposition 3.4. The map $v \in H_A \mapsto \ell_v \in \mathcal{D}'$ with $\ell_v(w) = (w, v)_A$ is continuous and injective, with a dense image in the strong dual $\mathcal{D}'$ of $\mathcal{D}$. One has $\mathcal{D}' = \cup_{i\geq 0} H_{A_i}$ (identifying H with its image in $\mathcal{D}'$), and the topology of $\mathcal{D}'$ is the inductive limit of the sequence of Hilbert spaces H_{A_i}, $i \in \mathbb{N}$. Moreover, $\mathcal{D}$ is dense in strong $\mathcal{D}'$.

Proof. Let $j_A : \mathcal{D} \to H_A$ be the continuous canonical injection from the Fréchet space $\mathcal{D}$ into H_A. The transposed map ${}^t j_A$ is the map $v \to \ell_v$ of the proposition and is injective, since $j_A(\mathcal{D})$ is dense in H_A. Now, $\mathcal{D}$ and H_A being reflexive spaces, one has ${}^{tt}j_A = j_A$, $\mathcal{D} = \mathcal{D}''$ (the second dual) and $H_A = H_A''$. Injectivity of ${}^{tt}j_A$ implies that ${}^t j_A\ (H_A')$ is dense in $\mathcal{D}'$ for weak topology $\sigma(\mathcal{D}', \mathcal{D}'')$. The weak closure of ${}^t j_A\ (H_A')$ coïncides with its strong closure, by the Minkowski theorem, namely ${}^t j_A\ (H_A')$ is dense in strong $\mathcal{D}'$. Taking A = Id, we find that the Hilbert space H, or (equivalently) its dense linear subset $\mathcal{D}$, is continuously embedded in $\mathcal{D}'$ with dense image. The formula $\mathcal{D}' = \cup_{i\geq 0} H_{A_i}$ is contained in proposition 3.3. Finally, $\mathcal{D}'$ strong is a bornological DF-space, and hence is the inductive limit of the system of normed spaces $\mathcal{D}'_B$ obtained from the closed bounded (resp., equicontinuous) disked subset B of $\mathcal{D}'$. Such B are described in proposition 3.3, thus proving our proposition.

Proposition 3.5. Let $\mathcal{A}$ be a $*$-algebra satisfying condition I_0, with natural domain $\mathcal{D}$. Let $A \in \mathcal{A}$, $A \geq$ Id such that all powers A^n, $n \in \mathbb{N}$ are essentially self-adjoint on $\mathcal{D}$, and $\mathcal{D}_A = \cap_{n\geq 0} \mathrm{Dom}(\bar{A}^n)$ the (Fréchet) space endowed with semi-norms $x \in \mathcal{D}_A \to \|\bar{A}_n x\|$, for $n \in \mathbb{N}$. Then $\mathcal{D}_A \supset \mathcal{D}$ and the natural injection from $\mathcal{D}$ into $\mathcal{D}_A$ is continuous for respective Fréchet topologies, and has a dense image.

Proof. Since $A^n \in \mathcal{A}$ for $n \geq 0$, formula $\mathcal{D} \subset \mathcal{D}_A$ is straightforward, as well as continuity of the canonical injection. From $\bar{A}^p\ \mathcal{D}_A = \mathcal{D}_A$ ($p \in \mathbb{N}$), by corollary 3.1, we find that A^p is essentially self-adjoint respectively on $\mathcal{D}_A$ and on $\mathcal{D}$. Thus $\mathcal{D}_A$ and $\mathcal{D}$ are core of $\mathrm{Dom}(\bar{A}^{p/2})$ for all $p \geq 0$, implying that $\mathcal{D}$ is dense in $\mathcal{D}_A$ for all seminorms $x \to \|\bar{A}^p x\|$, thus proving the proposition.

Proposition 3.6. Let $\mathcal{A}$ be the natural domain of a $*$-algebra and H the Hilbert space completion of $\mathcal{D}$. For every sequence x_n in $\mathcal{D}$ such that Ax_n is a weakly convergent sequence in H (i.e., for topology $\sigma(H, H')$) for all $A \in \mathcal{A}$, we have that the weak limit of the sequence x_n exists and is in $\mathcal{D}$.

This states condition (5) of [22]. In fact, for $\mathcal{A} = \cup_{i\geq 0} \mathcal{A}_{A_i}$, it is sufficient to check that $A_i x_n$ is weakly convergent for all $i \geq 0$.

Proof. From proposition 3.3, for every $L \in \mathcal{D}'$, we get that $L(x_n - x_m) = (A_i(x_n - x_m), \bar{A}_i\zeta)$ tends to zero as $n, m \to \infty$, hence x_n is a Cauchy sequence for $\sigma(\mathcal{D}, \mathcal{D}')$. Since $\mathcal{D}$ is reflexive, $\sigma(\mathcal{D}, \mathcal{D}') = \sigma(\mathcal{D}'', \mathcal{D}')$, and since $\|A_i x_n\|$ is bounded for every $i \geq 0$, by the Banach-Steinhaus theorem, the set of x_n is equicontinuous in $\mathcal{D}''$, hence admits a weak limit $x_0 \in \mathcal{D}$ for $\sigma(\mathcal{D}'', \mathcal{D}')$. Linear forms $x \in \mathcal{D} \to (x, y)$ with y given in H belong to $\mathcal{D}'$, so that x_0 is the limit for $\sigma(H, H')$ of the sequence x_n.

Corollary 3.2. Let $\mathcal{A}$ ultraweakly closed with condition II. If the Hilbert space H is separable, then the predual P_A of $\mathcal{A}$ is a separable Fréchet space.

Indeed, the map $\theta: \Sigma\, x \otimes y \in \mathcal{D} \hat{\otimes} \mathcal{D} \longmapsto \Sigma\, \omega_{x,y} \in P_A$ is a topological homomorphism from the (Fréchet space) projective tensor product of the Fréchet space $\mathcal{D}$ by itself, onto P_A and is an isomorphism when $\mathcal{A} = B(\mathcal{D}, \mathcal{D})$. Separability of $\mathcal{D} \hat{\otimes} \mathcal{D}$ is equivalent to that of $\mathcal{D}$, since $\mathcal{D}$ is isomorphic to a direct factor of $\mathcal{D} \hat{\otimes} \mathcal{D}$, and it suffices to apply proposition 3.7. Let $\varphi = \varphi^* \in P_A$. It is obvious that there exists $\varphi_1 \geq 0$, $\varphi_2 \geq 0$, $\varphi_1, \varphi_2 \in P_A$ such that $\varphi = \varphi_1 - \varphi_2$. Any compact subset of

$$(\mathcal{D} \hat{\otimes} \mathcal{D})/\mathcal{A}^\circ \approx P_A$$

(with $\mathcal{A}^\circ$ polar of $\mathcal{A}$ in duality $\langle \mathcal{D} \hat{\otimes} \mathcal{D}, B(\mathcal{D}, \mathcal{D})\rangle$ is the canonical image of a compact set of $\mathcal{D} \hat{\otimes} \mathcal{D}$, thus, for $\varphi = \varphi^*$ moving in a compact set of P, we can choose $\varphi_1 \geq 0$, $\varphi_2 \geq 0$ in such a way that φ_1 (resp. φ_2) moves in a positive compact set of P. In fact, for any sequence φ_n of the predual tending to zero in P, we can find φ_n^+ , φ_n^- , positive in P such that

$\varphi_n = \varphi_n^+ - \varphi_n^-$ with φ_n^+ (resp. φ_n^-) tending to zero in P: indeed, convergent sequences of $(\mathcal{D} \hat{\otimes} \mathcal{D})/A°$ are canonical image, via θ, of convergent sequences of $\mathcal{D} \hat{\otimes} \mathcal{D}$, and we may refer to remark 4 of [8].

Proposition 3.7. Let $\mathcal{D}$ be the natural domain of a $\star$-algebra A acting in the Hilbert space H. Then, the Fréchet space $\mathcal{D}$ is separable if and only if H is separable.

Proof. Since the topology of $\mathcal{D}$ is finer than the topology of H restricted to $\mathcal{D}$, the separability of $\mathcal{D}$ implies that of H. Let now A_i be a cofinal sequence in A^+, with $\|A_i x\| \geq \|x\|$ for all $x \in \mathcal{D}$ and $i \geq 0$. Every A_i defines the disked neighborhood $U_i = \{x \in \mathcal{D} : \|A_i x\| \leq 1\|\}$, the normed space $\mathcal{D}_{U_i}$ (equal as vector space to $\mathcal{D}$) with unit ball U_i and $\mathcal{D}_{U_i}$ can be identified with a topological subspace of H by the map $x \in \mathcal{D}_{U_i} \to A_i x \in H$. The separability of H implies the separability of normed spaces $\mathcal{D}_{U_i}$ and, $\mathcal{D}$ being the projective limit of the sequence of normed spaces $\mathcal{D}_{U_i}$, we get the separability of $\mathcal{D}$.

The following notion, introduced by von Neumann in [17], is often used in our proofs.

Definition 3.2. Let H be a Hilbert space and $\mathcal{D}$ be a dense linear subset of H. Let M be a von Neumann algebra acting in H. If a sequence $K_1, K_2, \ldots$ of closed linear subsets of H exists which has the following properties:

(i) $K_i \,\eta\, M$,

(ii) $K_1 \subset K_2 \subset \cdots \subset \mathcal{D}$,

(iii) $(\cup_i K_i)^\perp = 0$,

then $\mathcal{D}$ is called essentially dense.

Given a sequence (f_n) of measurable real functions on a measured space X with positive measure p, there exists, by Egoroff's theorem, a sequence $E_1, E_2, \ldots, E_n, \ldots$ of measurable sets in X such that $p(X - \cup_n E_n) = 0$ with the property that each f_n is bounded on each E_i. The role of such measurable sets is played, in our context, by projectors E of M such that $E(H) \subset \mathcal{D}$. For projectors E_1, E_2 in M with $E_1(H) \subset \mathcal{D}$, $E_2(H) \subset \mathcal{D}$ one clearly has $\sup(E_1, E_2)(H) \subset \mathcal{D}$ and $F(H) \subset \mathcal{D}$ for $F \leq E_1$, F projector. For a net β_α in $B(\mathcal{D}, \mathcal{D})$ with ρ-limit $\beta \in B(\mathcal{D}, \mathcal{D})$ and a given projector E with $E(H) \in \mathcal{D}$, we find from proposition 2.3, that β_α E × E are bounded operators in $L(H)$ converging in norm to β E × E ∈ $L(H)$. Lemma 16.2.2 and 16.2.3 of [17] lead to

Proposition 3.8. Let H be a Hilbert space and $\mathcal{D}$ be a dense linear subset of H. We assume that there exists a sequence $(A_i)_{i\in\mathbb{N}}$ of closeable operators $A_i \ \eta\ M$, $A_i \geq \mathrm{Id}$, $A_0 = \mathrm{Id}$ such that $\mathcal{D} = \cap_{i\geq 0} \mathrm{Dom}(\bar{A}_i)$. If the von Neumann algebra P generated by operators A_i^{-1} $i \geq 0$ is finite, then $\mathcal{D}$ is essentially dense.

We often use the proposition for abelian M. To be more complete, let M be a finite von Neumann algebra, and $X_0 = \mathrm{Id}$, X_1, X_2, ... be a sequence of closed operators with $X_i \eta M$ for $i \geq 0$. Let A be the space of all operators $p(X_1, X_1^+, X_2, X_2^+, \ldots)$ where $p = p(x_1, y_1, x_2, y_2, \ldots)$ moves in the set of all non commutative polynomials in variables x_1, y_1, x_2, y_2, ..., and $\mathcal{D}$ be the common part of domains of the closure of such operators; then, $\mathcal{D}$ is essentially dense in H, and A is a countably dominated $*$-algebra with natural domain $\mathcal{D}$, consisting of operators ηM. The details are left to the reader.

Lemma 3.3. Let $A = \cup_{i\geq 0} A_{A_i}$ (with $A_0 = \mathrm{Id}$) be an ultraweakly closed subspace of $B(\mathcal{D}, \mathcal{D})$ with condition II, and let $\mathcal{D}$ be essentially dense. The set B of T in A_{id} such that TA_i (already defined on $\mathcal{D}$) extends continuously to a bounded operator in H, for all $i \geq 0$, is a left ideal in A_{id}, Hilbert ultraweakly dense in A_{id}, and $\sigma(\mathcal{D} \hat{\otimes} \mathcal{D})$ dense in A.

Proof. For T ∈ B, S ∈ A_{id}, and the formula $\|STA_i x\| \leq \|S\|\ \|TA_i x\|$ with $x \in \mathcal{D}$, we find ST ∈ B. Moveover, the formula

$$(TA_i x, y) = (A_i x, T^* y) = (x, (TA_i)^+ y) = (x, \bar{A}_i T^* y)$$

for $x \in \mathrm{Dom}(\bar{A}_i)$, $y \in H$ shows that $\bar{A}_i T^*$ is bounded on H hence $T^*(H) \subset \mathrm{Dom}(\bar{A}_i)$ implies $T^*(H) \subset \mathcal{D}$, since this holds for any $i \geq 0$, thus $\bar{A}_i T^* = A_i T^*$ on $\mathcal{D}$. Due to $A_i T^*(H) \subset \mathcal{D}$ we see from the closed graph theorem that, for every integer $j \geq 0$, $A_i A_j T^*$ is continuous from the Hilbert space H into the Fréchet space $\mathcal{D}$, thus $A_i (TA_j)^*$ is bounded for every $i \geq 0$, as well as its adjoint $(TA_j)A_i$, i.e., $TA_j \in \mathcal{B}$. The proof is similar for A_j^{-1}. Now let $T \in \mathcal{A}_{id}$ and p_n be an increasing sequence of projectors in $\mathcal{A}_{id}$, increasing to Id, with $p_n(H) \subset \mathcal{D}$. Clearly, T is the Hilbert strong limit of the bounded sequence Tp_n and we are lead to show that $Tp_n \in \mathcal{B}$: but, for $i \geq 0$, $A_i p_n$ is a bounded operator with domain H, whose adjoint agree on $\mathcal{D}$ with the bounded operator $p_n A_i$, so that $(Tp_n)A_i = T(p_n A_i)$ must be bounded. Finally, $\mathcal{B}$ is $\sigma(H \hat{\otimes} H)$ dense in the von Neumann algebra $\mathcal{A}_{id}$, therefore dense in $\mathcal{A}$ for $\sigma(\mathcal{D} \hat{\otimes} \mathcal{D})$-topology by theorem 1.2.

Lemma 3.4. Let $\mathcal{A}$, $\mathcal{B}$, $\mathcal{D}$ of lemma 3.3, and $\mathcal{C}$ be the set of B in $\mathcal{A}_{id}$ such that sesquilinear forms $(x, y) \in \mathcal{D}\times\mathcal{D} \to (BA_i x, B_i y)$ and $(B^* A_i x, A_i y)$ extend continuously to $H \times H$ for all $i \geq 0$. Then:

1°/ $\mathcal{C}$ is an involutive algebra of bounded operators, contained in $L(\mathcal{D})$, stable under $\beta \to \beta A_i \times A_i$, $\beta \to \beta A_i^{-1} \times A_i^{-1}$, and $\mathcal{C} \subset \mathcal{B} \cap \mathcal{B}^*$.

2°/ The von Neumann algebra generated by $\mathcal{C}$ is $\mathcal{A}_{id}$.

Proof. For $B \in \mathcal{C}$ and $i \geq 0$, the corresponding form $BA_i \times A_i$ is associated to the bounded operator $\bar{A}_i BA_i$, defined on $\mathcal{D}$. Since $\bar{A}_i^{-1}$ is bounded and $\mathcal{C} = \mathcal{C}^*$, we find that $\bar{A}_i^{-1} \bar{A}_i BA_i = BA_i$ and $B^* A_i$ are bounded operators, hence $B(H) \subset \mathcal{D}$, $B^*(H) \subset \mathcal{D}$ by the proof of lemma 3.3. Let $B_1 = B_1^* \in \mathcal{C}$, $B_2 = B_2^* \in \mathcal{C}$. Using

$$(B_1 B_2 A_j x, A_j y) = (B_2 A_j x, B_1 A_j y) =$$

$$= \frac{1}{4} \sum_{\alpha=1}^{4} ((B_1 + i^\alpha B_2)A_j x, (B_1 + i^\alpha B_2)A_j y)$$

with $j \geq 0$, we see that $B_1 B_2 \in \mathcal{C}$ as soon as $BA_i \times BA_i$ is a bounded operator for $B \in \mathcal{C}$, which follows from the boundedness of BA_i. For B in $\mathcal{C}$ and $j \geq 0$, $BA_j \times A_j$ is a bounded operator of $\mathcal{A}_{id}$, since it commutes with the von Neumann algebra $\mathcal{A}'_{id}$. For every $i \geq 0$, there exists a constant $M < +\infty$ and $k \in \mathbb{N}$ such that $\|A_i A_j x\| \leq M\|A_k x\|$ for all $x \in \mathcal{D}$, implying that $A_i A_j = CA_k$ with $C \in \mathcal{A}_{id}$, $C\mathcal{D} = \mathcal{D}$, $C^*\mathcal{D} = \mathcal{D}$. Thus $BA_j A_i \times A_j A_i = BA_k C^* \times A_k C^*$ is in $\mathcal{C}$, since $BA_k \times A_k \in \mathcal{C}$. The proof is similar for $BA_j^{-1} \times A_j^{-1}$. Now let $T = T^* \in \mathcal{A}_{id}$ and p_n be an increasing sequence of projectors in A_{id}, with $p_n(H) \subset \mathcal{D}$, $p_n \to 1$. It is easily seen that T is the Hilbert ultraweak limit of the sequence $p_n\ T\ p_n$, so that we need to show that $Tp_n \times p_n \in \mathcal{C}$. Taking $i \geq 0$, we see that $T\ p_n A_i \times p_n A_i$ is a bounded operator, since $(A_i p_n)$ and T are so.

Theorem 3.1. Let A be an ultraweakly closed subspace of $B(\mathcal{D}, \mathcal{D})$ satisfying condition II, with $\mathcal{D}$ essentially dense. There exists a $*$-algebra A_0 with natural domain $\mathcal{D}$ satisfying condition I, ultraweakly dense in A.

It suffices to choose for A_0 the $*$-algebra generated by all A_i, A_i^{-1} and $\mathcal{C}$ of lemma 3.4.

Proposition 3.9. Let $A = \cup_i A_{A_i}$ be a $*$-algebra with natural domain $\mathcal{D}$ and M be a finite von Neumann algebra such that any T in A is affiliated to M.

1°/ Every linear subset $\mathcal{D}_1$ of $\mathcal{D}$, dense in the Hilbert space H and stable under M' is dense in the Fréchet space $\mathcal{D}$.

2°/ For every $T \geq \mathrm{Id}$ of A, there exists a dense linear subset $\mathcal{D}_T$ of the Fréchet space $\mathcal{D}$ such that $T(\mathcal{D}_T) = \mathcal{D}_T$. When T commute on $\mathcal{D}$ with all A_i, one can choose $\mathcal{D}_T = \mathcal{D}$.

Proof. For T in A, let $\bar{T}$ (resp. $\bar{T}_1$) be the closure of the restriction of T to $\mathcal{D}$ (resp. to $\mathcal{D}_1$). It is easily seen that $M'\mathcal{D} \subset \mathcal{D}$ and from $M'\mathcal{D}_1 \subset \mathcal{D}_1$, we get that the operators T, defined respectively on $\mathcal{D}$ and $\mathcal{D}_1$, are affiliated to M, as well as $\bar{T}$ and $\bar{T}_1$, by lemma 1.4. Since $\bar{T}$ extends $\bar{T}_1$ we get, by theorem XV of [17], that $\mathcal{D}_1$ is a core of $\mathrm{Dom}(\bar{T})$, thus taking T in a cofinal sequence of A^+, we find that $\mathcal{D}_1$ is dense in the Fréchet space $\mathcal{D}$. Now let $T \geq \mathrm{Id}$, $T \in A$. By lemma 16.2.3 [17], the set $\mathcal{D}_T = \cap_{k \in \mathbb{Z}} \{v \in \mathcal{D} | \bar{A}^k\ v \in \mathcal{D}\}$ is essentially dense and one can check that $T(\mathcal{D}_T) = \mathcal{D}_T$, noting that A is essentially selfadjoint on $\mathcal{D}$. For B in M', $v \in \mathcal{D}$ and formula $\bar{A}^{-k}\ Bv = B\ \bar{A}^{-k}\ v$ for $k \geq 0$, we find $Bv \in \mathcal{D}_T$ thus, by 1°/, $\mathcal{D}_T$ is dense in the Fréchet space $\mathcal{D}$. The last assertion is straightforward.

Proposition 3.10. Let $A = \cup_i A_{A_i}$ be an ultraweakly closed space with condition II and natural domain $\mathcal{D}$. Every linear subset $\mathcal{D}_1$ dense in the pre-Hilbert space $\mathcal{D}$ and stable by the von Neumann algebra generated by all A_i^{-1} must be dense in the Fréchet space $\mathcal{D}$.

Proof. For every $i \geq 0$, the closure $\bar{A}_i$ of A_i is a self-adjoint operator $\geq \mathrm{Id}$, and the corresponding one parameter group $t \to e^{itA}$ lies in P. Clearly $\mathcal{D}_1 \subset \mathcal{D} \subset \mathrm{Dom}(\bar{A}_i)$ and stability of $\mathcal{D}_1$ under all e^{itA} implies that $\mathcal{D}_1$ is a core of $\mathrm{Dom}(\bar{A}_i)$. Since this holds for every $i \geq 0$, we get the proposition.

Proposition 3.11. Let A be a $*$-algebra satisfying condition I, with natural domain $\mathcal{D}$, and $\zeta \in \mathcal{D}$ a cyclic vector for the involutive algebra A_{id}. Then, $A_{id}\zeta$ is a dense linear subspace of the Fréchet space $\mathcal{D}$.

Proof. Let $f \in \mathcal{D}'$ equal to zero on $A_{id}\zeta$. There exists by proposition 3.3, an integer $i \geq 0$ and $y \in H$ such that $f(v) = (A_i v, y)$ for all $v \in \mathcal{D}$. One has, for $B \in A_{id}$, $(A_i B\zeta, y) = 0$, and taking B of the form $B = A_i^{-1}C$ with C moving in A_{id}, we find $(C\zeta, y) = 0$ for $C \in A_{id}$ hence $y = 0$ implying $f = 0$. It follows the density of $A_{id}\zeta$ in $\mathcal{D}$.

We will need the preliminary

Proposition 3.12. Let M be a von Neumann algebra with commutant M', and $\Delta \geq \mathrm{Id}$ (resp. $\nabla \geq \mathrm{Id}$) be a self-adjoint operator affiliated with M (resp. M').

Let $\mathcal{D}_\Delta = \cap_{n\geq 0} \Delta^{-n}(H)$, $\mathcal{D}_\nabla = \cap_{n\geq 0} \nabla^{-n}(H)$, $\Delta\nabla$ (or $\nabla\Delta$) be the inverse of the (closed) operator $\Delta^{-1}\nabla^{-1} = \nabla^{-1}\Delta^{-1}$ and $\mathcal{D}_{\Delta\nabla} = \cap_{n\geq 0} (\nabla^{-1}\Delta^{-1})^n(H)$.

1°/ One has $\Delta(\mathcal{D}_\Delta) = \mathcal{D}_\Delta$, $\nabla(\mathcal{D}_\nabla) = \mathcal{D}_\nabla$, $\Delta\nabla$ is self adjoint and $\geq \mathrm{Id}$, $\Delta\nabla(\mathcal{D}_{\Delta\nabla}) = \mathcal{D}_{\Delta\nabla}$, $\Delta(\mathcal{D}_{\Delta\nabla}) = \mathcal{D}_{\Delta\nabla}$, $\nabla(\mathcal{D}_{\Delta\nabla}) = \mathcal{D}_{\Delta\nabla}$.

2°/ $\mathcal{D}_{\Delta\nabla} = \{x \in \mathcal{D}_\Delta \cap \mathcal{D}_\nabla \mid \Delta^n x \in \mathcal{D}_\nabla \text{ for all } n \geq 0\}$

$= \{x \in \mathcal{D}_\Delta \cap \mathcal{D}_\nabla \mid \nabla^n x \in \mathcal{D}_\Delta \text{ for all } n \geq 0\}$

$= \{x \in \mathcal{D}_\Delta \cap \mathcal{D}_\nabla \mid \Delta^n x \in \mathcal{D}_\nabla \text{ and } \nabla^n x \in \mathcal{D}_\Delta \text{ for all } n \geq 0\}$.

3°/ $\mathcal{D}_{\Delta\nabla}$ is a dense linear subset of the Fréchet space $\mathcal{D}_\Delta$ (resp. $\mathcal{D}_\nabla$).

Due to symmetric roles of Δ and ∇, $\mathcal{D}_{\Delta\nabla} = \mathcal{D}_{\nabla\Delta}$.
The reader will check that, for $k \geq 0$, the product $\nabla^k \cdot \Delta^k$ of the self-adjoint operator Δ^k and ∇^k is a densely defined essentially self-ajoint operator in the Hilbert space H, and its closure $\overline{\nabla^k \cdot \Delta^k}$ is exactly $(((\Delta\nabla)^{-k})^{-1}$ and agrees with $\overline{\Delta\nabla}^k$ (use theorem XV in [17]).

Proof. For x in the Hilbert space H, one has

$$0 \leq (\Delta^{-1}\nabla^{-1}x, x) \leq \|\nabla^{-1}x\| \, \|\Delta^{-1}x\| \leq \|x\|^2$$

and, since $\Delta^{-1}\nabla^{-1}$ is injective in H, $\Delta\nabla$ is $\geq$ Id and must be a self-adjoint operator being the inverse of the bounded self-adjoint operator $\nabla^{-1}\Delta^{-1}$; thus, $(\nabla\Delta)^{-1} = \nabla^{-1}\Delta^{-1}$.
Noting that $\mathcal{D}_\Delta$ is the space of differentiable vectors for the one-parameter group $t \to e^{it\Delta}$, we get $\Delta(\mathcal{D}_\Delta) = \mathcal{D}_\Delta$; similarly, $\nabla(\mathcal{D}_\nabla) = \mathcal{D}_\nabla$, $(\Delta\nabla)(\mathcal{D}_{\Delta\nabla}) = \mathcal{D}_{\Delta\nabla}$, $((\Delta\nabla)^{-1})^n = \nabla^{-n}\Delta^{-n}$, leading to $(\Delta\nabla)^{-1}(\mathcal{D}_{\Delta\nabla}) = \mathcal{D}_{\Delta\nabla}$. Inclusion $\Delta^{-1}(\mathcal{D}_{\Delta\nabla}) \subset \mathcal{D}_{\Delta\nabla}$ follows from

$$\Delta^{-1}(\Delta^{-1}\nabla^{-1}(H) \cap \Delta^{-2}\nabla^{-2}(H) \cap \cdots) \subset$$

$$\subset \Delta^{-1}\nabla^{-1}\Delta^{-1}(H) \cap \Delta^{-2}\nabla^{-2}\Delta^{-1}(H) \cap \cdots$$

$$\subset \Delta^{-1}\nabla^{-1}(H) \cap \Delta^{-2}\nabla^{-2}(H) \cap \cdots = \mathcal{D}_{\Delta\nabla} ,$$

due to $\Delta^{-1}(H) \subset H$, and, similarly, $\nabla^{-1}\mathcal{D}_{\Delta\nabla} \subset \mathcal{D}_{\Delta\nabla}$.
As $\Delta^{-1}\nabla^{-1}(\mathcal{D}_{\Delta\nabla}) = \mathcal{D}_{\Delta\nabla}$ we are led to $\Delta^{-1}(\mathcal{D}_{\Delta\nabla}) = \mathcal{D}_{\Delta\nabla}$, due to injectivity of ∇^{-1}, Δ^{-1}. Composing each side by Δ, and noting that $\mathcal{D}_{\Delta\nabla} \subset \mathcal{D}_\Delta$, we get $\Delta(\mathcal{D}_{\Delta\nabla}) = \mathcal{D}_{\Delta\nabla}$; hence 1°/.

For $x \in \mathcal{D}_{\Delta\nabla}$, one has, for all $n \geq 0$, $\Delta^n x \in \mathcal{D}_{\Delta\nabla} \subset \mathcal{D}_\nabla$ and (or) $\nabla^n x \in \mathcal{D}_{\Delta\nabla} \subset \mathcal{D}_\Delta$ due to 1°/, and $x \in \cap_n \Delta^{-n}\nabla^{-n}(H) \subset \cap_n \Delta^{-n}(H) = \mathcal{D}_\Delta$ (resp. $\cap_n \nabla^{-n}(H) = \mathcal{D}_\nabla$), i.e., $x \in \mathcal{D}_\Delta \cap \mathcal{D}_\nabla$. Conversely, if $x \in \mathcal{D}_\Delta \cap \mathcal{D}_\nabla$ satisfies, for all $n \geq 0$, $\Delta^n x \in \mathcal{D}_\nabla$ and (or) $\nabla^n x \in \mathcal{D}_\Delta$, we get $x \in \Delta^{-n}(\mathcal{D}_\nabla) \subset \Delta^{-n}\nabla^{-n}(H) = (\Delta\nabla)^{-n}(H)$, thus $x \in \cap_{n\geq 0}(\Delta\nabla)^{-n}(H) = \mathcal{D}_{\Delta\nabla}$; hence 2°/ due to symmetric roles of Δ and ∇.
As $e^{it\Delta}\nabla^{-k} = \nabla^{-k}e^{it\Delta}$ for $k \geq 0$, we first get

$$e^{it\Delta}(\mathcal{D}_\nabla) = e^{it\Delta}(\bigcap_{k\geq 0} \nabla^{-k}(H)) \subset \bigcap_{k\geq 0} \nabla^{-k}(H) = \mathcal{D}_\nabla.$$

Since differentiable vectors for Δ^n, $n \geq 1$, agree with those for Δ, we find, for all $n \geq 0$, that $e^{it\Delta^n}(\mathcal{D}_\nabla) \subset \mathcal{D}_\nabla$. Formula $\Delta^{-1}\nabla^{-1} = \nabla^{-1}\Delta^{-1}$ implies $e^{it\Delta}e^{is\nabla} = e^{is\nabla}e^{it\Delta}$ for $t \in \mathbb{R}$, $s \in \mathbb{R}$, hence $e^{it\Delta}\nabla^k \subset \nabla^k e^{it\Delta}$ for $k \geq 0$. Given x in $\mathcal{D}_\Delta \cap \mathcal{D}_\nabla$ with $\nabla^k x \in \mathcal{D}_\Delta$ for $k \geq 0$, one has, from $e^{it\Delta}(\mathcal{D}_\Delta) = \mathcal{D}_\Delta$, $e^{it\Delta}x \in \mathcal{D}_\Delta \cap \mathcal{D}_\nabla$ and $\nabla^k e^{it\Delta}x = e^{it\Delta}\nabla^k x \in e^{it\Delta}(\mathcal{D}_\Delta) = \mathcal{D}_\Delta$; this establishes $e^{it\Delta}(\mathcal{D}_{\nabla\Delta}) \subset \mathcal{D}_{\Delta\nabla}$. Replacing (as above) Δ by Δ^n. ($\mathcal{D}_\Delta = \mathcal{D}_{\Delta^n}$), we may use the proof of proposition 3.10 to conclude 3°/, since $\mathcal{D}_{\Delta\nabla} \subset \mathcal{D}_\Delta$ is stable under all unitary groups $t \mapsto e^{it\Delta^n}x$, $n \geq 0$.

Being concerned mainly with spaces A with a cofinal abelian sequence, we first show, with the help of [17],

Lemma 3.5. Let N be an abelian von Neumann algebra, and $\Delta_1, \Delta_2, \ldots$ be a sequence of self-adjoint operators $\geq$ Id, such that $\Delta_i^{-1} \in N$ for all i.

1°/ $\mathcal{D} \equiv \bigcap_{i\geq 0, k\geq 0} \Delta_i^{-k}\Delta_{i-1}^{-k} \cdots \Delta_1^{-k}(H)$ is an essentially dense domain for N and is the natural domain of the $*$-algebra generated 'formally' by all Δ_i^k i, $k \geq 0$; one has, for $k \in \mathbb{Z}$ and $i \geq 0$, $\Delta_i^k\mathcal{D} = \mathcal{D}$. If $0 \leq \Delta_{n+1}^{-1} \leq \Delta_n^{-1} \leq$ Id for all n, one has $\mathcal{D} = \bigcap_{i\geq 0\ k\geq 0} \Delta_i^{-k}(H)$.

2°/ For a subset $J \subset \mathbb{N}$, let $\mathcal{D}_J$ be the domain constructed in 1°/ corresponding to the subsequence Δ_j with $j \in J$. Then the Fréchet space $\mathcal{D}_J$ contains the (Fréchet) space $\mathcal{D}$ as a dense linear subset.

We note that $\Delta_n^{-1} \in N$ is equivalent to $\Delta_n \eta N$.

Proof. We first show that one may assume that $0 \leq \Delta_{n+1}^{-1} \leq \Delta_n^{-1} \leq$ Id, for all $n \geq 0$, in this proof. Since $\Delta_\alpha^{-1}\Delta_{\alpha-1}^{-1} \cdots \Delta_1^{-1}$ is a self-adjoint positive and injective operator with norm ≤ 1, as N is abelian, its inverse A_α, for $\alpha \in \mathbb{N}$, makes sense and is self-adjoint $\geq$ Id (with domain equal to $\Delta_\alpha^{-1} \cdots \Delta_1^{-1}(H))$. The inverse of $(A_\alpha)^n = A_\alpha^n$ is the n-th power of the A_α^{-1} (by the spectral theorem, for example), i.e., $A_\alpha^n = (\Delta_\alpha^{-n} \cdots \Delta_1^{-n})^{-1}$, for $n \geq 0$. For $n \geq 0$, $\alpha \geq 0$, one has $\Delta_\alpha^{-n} \leq \Delta_\alpha^{-n} \cdots \Delta_1^{-n} \leq \mathrm{Id}$ due to $\Delta_\alpha^{-1} \leq \Delta_j^{-1}$ for $0 \leq j \leq \alpha$, thus

$$\bigcap_{n\geq 0} \bigcap_{k\geq 0} \Delta_\alpha^{-n}(H) \subset \bigcap_{n\geq 0} \bigcap_{k\geq 0} \Delta_\alpha^{-n} \cdots \Delta_1^{-n}(H) \; ;$$

noting here that $0 \leq B \leq C \leq \mathrm{Id}$ and $BC = CB$ implies that $B^{\frac{1}{2}}(H) \subset C^{\frac{1}{2}}(H)$, leading to $B(H) \subset C(H)$. Due to $\Delta_{\alpha-1}^{-n} \cdots \Delta_1^{-n}(H) \subset H$, we get $A_\alpha^{-n}(H) \subset \Delta_\alpha^{-n}(H)$ for all $n \geq 0$, and this leads, together with the preceding inclusion, to

$$\bigcap_{n\geq 0} \bigcap_{k\geq 0} \Delta_\alpha^{-n}(H) = \bigcap_{n\geq 0} \bigcap_{k\geq 0} A_\alpha^{-n}(H).$$

Thus our proposition is reduced to the situation $0 \leq \Delta_{n+1}^{-1} \leq \Delta_n^{-1} \leq \mathrm{Id}$ for $n \geq 0$. For $p \in \mathbb{N}$, let $\mathcal{D}_p = \cap_{n\geq 0} \Delta_p^{-n}(H)$. From $\mathcal{D} = \cap_p \cap_k \Delta_p^{-k}(H)$ we find $\mathcal{D} = \cap_p \mathcal{D}_p$. Let k be a fixed integer. We will show that $\Delta_k^{-1}\mathcal{D} = \mathcal{D}$, $\Delta^k \mathcal{D} = \mathcal{D}$. For $j \geq k$, one has $\Delta_j^{-1} \in N'$ (as $N' \supset N$), thus, by proposition 3.12, $\Delta_k^\varepsilon(\mathcal{D}_{jk}) = \mathcal{D}_{jk}$ for $\varepsilon \in \mathbb{Z}$, where $\mathcal{D}_{jk} = \cap_{n\geq 0} \Delta_j^{-n} \Delta_k^{-n}(H)$. For $n \geq 1$, $\Delta_j^{-1} \leq \Delta_k^{-1} \leq \mathrm{Id}$ implies $\Delta_j^{-n} \leq \Delta_k^{-n} \leq \mathrm{Id}$ since N is abelian, hence $\Delta_j^{-2n} \leq \Delta_k^{-n}\Delta_j^{-n} \leq \Delta_j^{-n}$ leads to $\mathcal{D}_{kj} = \cap_{n\geq 1} \Delta_j^{-n}(H) = \mathcal{D}_j$. The same calculation shows that $\mathcal{D} = \cap_{p\geq k} \mathcal{D}_p$. Thus $\Delta_k^\varepsilon(\mathcal{D}_j) = \mathcal{D}_j$ for $\varepsilon = \pm 1$ implies $\Delta_k^\varepsilon(\cap_{p\geq k}\mathcal{D}_p) \subset \cap_{p\geq k} \Delta_k^\varepsilon(\mathcal{D}_p) \subset \cap_{p\geq k}\mathcal{D}_p = \mathcal{D}$ and, from injectivity of Δ_k^ε, we get $\Delta_k^\varepsilon \mathcal{D} = \mathcal{D}$ for $\varepsilon = \pm 1$, hence, for $\varepsilon \in \mathbb{Z}$. Each $\mathcal{D}_p$ is essentially dense relative to N, as well as $\mathcal{D} = \cap_p \mathcal{D}_p$ by [17] . As $\mathcal{D}$ is stable under N, we find, by proposition 3.10, that $\mathcal{D}$ is dense in the Fréchet space $\mathcal{D}_J$, i.e., 2°/.

The orientation of our analysis will be motivated by

<u>Proposition 3.13</u>. Let M be a von Neumann algebra with commutant M', and Δ_n (resp. ∇_n) be a sequence of self-adjoint operators $\geq \mathrm{Id}$ with commuting inverses satisfying

$0 \le \Delta_{n+1}^{-1} \le \Delta_n^{-1} \le \mathrm{Id}$ (resp. $0 \le \nabla_{n+1}^{-1} \le \nabla_n^{-1} \le \mathrm{Id}$) and $\Delta_n^{-1} \in M$ (resp. $\nabla_n^{-1} \in M'$) for all $n \ge 0$. We put $\mathcal{D}_{(\Delta)} = \cap_{n\ge 0}\ {}_{k\ge 0}\Delta_k^{-n}(H)$, $\mathcal{D}_{(\nabla)} = \cap_{n\ge 0}\ {}_{k\ge 0}\ \Delta_k^{-n}(H)$, and $\mathcal{D}_{(\Delta\nabla)} = \cap_{n\ge 0}\ {}_{k\ge 0}\Delta_k^{-n}\nabla_k^{-n}(H)$. Then

1°/ The Fréchet space $\mathcal{D}_{(\Delta)}$ (resp. $\mathcal{D}_{(\nabla)}$) is an essentially dense domain with respect to M (resp. M') and, for $i \ge 0$, $k \in \mathbb{Z}$, $\Delta_i^k\ (\mathcal{D}_{(\Delta)}) = \mathcal{D}_{(\Delta)}$ (resp. $\nabla_i^k(\mathcal{D}_{(\nabla)} = \mathcal{D}_{(\nabla)})$.

2°/ The Fréchet space $\mathcal{D}_{(\Delta\nabla)}$ is a dense linear subset of the Fréchet space $\mathcal{D}_{(\Delta)}$ (resp. $\mathcal{D}_{(\nabla)}$), and, for $i \ge 0$, $k \in \mathbb{Z}$, $\nabla_i^k(\mathcal{D}_{(\Delta\nabla)}) = \mathcal{D}_{(\Delta\nabla)}$ $\Delta_i^k(\mathcal{D}_{(\Delta\nabla)}) = \mathcal{D}_{(\Delta\nabla)}$.

3°/ The identity (or restriction) map from $B(\mathcal{D}_{(\Delta)}, \mathcal{D}_{(\Delta)})$ (resp. $B(\mathcal{D}_{(\nabla)}, \mathcal{D}_{(\nabla)})$) into $B(\mathcal{D}_{(\Delta\nabla)}, \mathcal{D}_{(\Delta\nabla)})$ is injective and continuous.

The subscript (Δ) (resp. (∇),...) in $\mathcal{D}_{(\Delta)}$ (resp. $\mathcal{D}_{(\nabla)}$) recalls that $\mathcal{D}_{(\Delta)}$ is constructed from a sequence (Δ).

The map of 3°/ sends the continuous sesquilinear form $\beta \in B(\mathcal{D}_{(\Delta)}, \mathcal{D}_{(\Delta)})$ or $B(\mathcal{D}_{(\nabla)}, \mathcal{D}_{(\nabla)})$ onto β viewed as a continuous sesquilinear form on $\mathcal{D}_{(\Delta\nabla)} \times \mathcal{D}_{(\Delta\nabla)}$. Of course, for a space $A \subset B(\mathcal{D}_{(\Delta)}, \mathcal{D}_{(\Delta)})$ with A ultraweakly closed relative to $\mathcal{D}_{(\Delta\nabla)} \hat{\otimes} \mathcal{D}_{(\Delta\nabla)}$, implies A ultraweakly closed relatively to $\mathcal{D}_{(\Delta)} \hat{\otimes} \mathcal{D}_{(\Delta)}$, the converse being, in general, false.

For completeness, we recall that $B(\mathcal{D}_{(\Delta)}, \mathcal{D}_{(\Delta)})$, $B(\mathcal{D}_{(\nabla)}, \mathcal{D}_{(\nabla)})$, $B(\mathcal{D}_{(\Delta\nabla)}, \mathcal{D}_{(\Delta\nabla)})$ is the space of continuous sesquilinear forms on the Fréchet space $\mathcal{D}_{(\Delta)} \times \mathcal{D}_{(\Delta)}$, $\mathcal{D}_{(\nabla)} \times \mathcal{D}_{(\nabla)}$, $\mathcal{D}_{(\Delta\nabla)} \times \mathcal{D}_{(\Delta\nabla)}$.

<u>Proof</u>. In lemma 3.5, we take for N the von Neumann algebra generated by all Δ_n^{-1}, ∇_n^{-1}, $n \ge 0$, leading immediately to $\Delta_i^k(\mathcal{D}_{(\Delta)}) = \mathcal{D}_{(\Delta)}$ and $\nabla_i^k(\mathcal{D}_\nabla) = \mathcal{D}_\nabla$ for $k \in \mathbb{Z}$, hence 1°/. The density property in 2°/ follows from proposition 3.10, for example. By lemma 3.5, $(\Delta\nabla)_i^{-1}\ (\mathcal{D}_{(\Delta\nabla)}) = \mathcal{D}_{(\Delta\nabla)}$ and, obviously, $(\Delta\nabla)_i^{-1} = \Delta_i^{-1}\nabla_i^{-1}$ since $(\Delta\nabla)_i$ is, by definition, the inverse of $\Delta_i^{-1}\nabla_i^{-1}$. One has

$$\Delta_i^{-1}(\mathcal{D}_{(\Delta\nabla)}) \subset \bigcap_{n\geq 0} \Delta_i^{-1}\Delta_k^{-n}\nabla_k^{-n}(H)$$

$$\subset \bigcap_{n\geq 0} \Delta_k^{-n}\nabla_k^{-n}\Delta_i^{-1}(H) \subset \bigcap_{n\geq 0} \Delta_k^{-n}\nabla_k^{-n}(H) = \mathcal{D}_{(\Delta\nabla)}$$

and, similarly, $\nabla_i^{-1}(\mathcal{D}_{(\Delta\nabla)}) \subset \mathcal{D}_{(\Delta\nabla)}$, leading to $\Delta_i^{-1}(\mathcal{D}_{(\Delta\nabla)} = \mathcal{D}_{(\Delta\nabla)}$ $\nabla_i^{-1}(\mathcal{D}_{(\Delta\nabla)}) = \mathcal{D}_{(\Delta\nabla)}$, hence 2°/, (from $\mathcal{D}_{(\Delta\nabla)} \subset \mathcal{D}_{(\Delta)} \cap \mathcal{D}_{(\nabla)}$, it also follows that $(\Delta\nabla)_i = \Delta_i\nabla_i$ on $\mathcal{D}_{(\Delta\nabla)}$. The proof of the third assertion is straightforward.

The reader can check

<u>Proposition 3.14</u>. $\mathcal{D}_{\nabla\Delta}$ is a Schwartz (resp. nuclear) space for its Fréchet topology iff there exists a projector E in the Hilbert space H commuting with Δ^{-1} , ∇^{-1}, and $\Delta^{-1}E$, $\nabla^{-1}(1-E)$ being compact (resp. nuclear) operators in the Hilbert space E(H) and (1-E)(H), respectively.

Thus, $\mathcal{D}_{\nabla\Delta}$ is a Schwartz (resp. nuclear) space as soon as $\mathcal{D}_{\Delta}$ or $\mathcal{D}_{\nabla}$ has the same property.

One now has to use a notion of tensor product concerning domains. This notion corresponds to the complete Hilbert tensor product of Hilbert spaces in cases where the domains are Hilbert spaces.
Let H_1 (resp. H_2) be a Hilbert space, and $H = H_1 \otimes H_2$ be the Hilbert complete tensor product. Let A_1 (resp. A_2) be a space with condition II acting in the Hilbert space H_1 (resp. H_2) with cofinal abelian sequence $(A_1)_i$ (resp. $(A_2)_i$) and $\mathcal{D}_1 = \bigcap_{i\geq 0} (A_1)_i^{-1}(H_1)$ (resp. $\mathcal{D}_2 = \ldots$) be its natural domain. Let $\mathcal{B}_1$ (resp. $\mathcal{B}_2$) be the abelian $\star$-algebra generated by all $(A_1)_i^{\varepsilon}$ (resp. $(A_2)_i^{\varepsilon}$), $i \in \mathbb{N}$ $\varepsilon = \pm 1$, with natural domain $\mathcal{D}_1$ (resp. $\mathcal{D}_2$). Since the description of $\mathcal{D}_1$(resp. $\mathcal{D}_2$) depends only on $\mathcal{B}_1$ (resp. $\mathcal{B}_2$), one may start directly from $\mathcal{B}_1$, $\mathcal{B}_2$ in order to define this tensor product.
For $X_1 = X_1^* \in \mathcal{B}_1$, $X_2 = X_2^* \in \mathcal{B}_2$, the linear operator $X_1 \otimes 1 + 1 \otimes X_2$ is essentially self-adjoint on the domain

$\mathcal{D}_1 \otimes \mathcal{D}_2$ (algebraic tensor product), thus we define $\mathcal{D}_1 \bar{\otimes} \mathcal{D}_2$ to be

$$\bigcap_{\substack{x_1 \in A_1 \\ x_2 \in A_2}} \mathrm{Dom}(\overline{x_1 \otimes 1 + 1 \otimes x_2}) =$$

$$= \bigcap_{i \geq 0} \Big((A_1)_i \otimes 1 + 1 \otimes (A_2)_i\Big)^{-1} (H_1 \otimes H_2),$$

a Fréchet space for the sequence of semi-norms

$$v \in \mathcal{D}_1 \bar{\otimes} \mathcal{D}_2 \longmapsto \|(A_1)_i \otimes 1 + 1 \otimes (A_2)_i)v\| .$$

Alternatively

<u>Definition 3.2</u>. $\mathcal{D}_1 \bar{\otimes} \mathcal{D}_2$ is the <u>natural</u> domain of the $*$-algebra $\mathcal{B}_1 \otimes \mathcal{B}_2$ generated linearly by all operators $S_1 \otimes S_2$ - defined on $\mathcal{D}_1 \otimes \mathcal{D}_2$ - with S_1, S_2 moving in $\mathcal{B}_1$, $\mathcal{B}_2$.

The domain $\mathcal{D}$ (i.e., $\mathcal{D} = \mathcal{D}_1$ or $\mathcal{D}_2$) is of type $\mathcal{D} = \cap_{n\geq 0} A^{-n}(H)$ iff $\mathcal{D}$ coïncide topologically with the Fréchet space $C^\infty(e^{itA})$ of differentiable vectors for the unitary one-parameter group $t \longmapsto e^{itA}$. Thus, $\mathcal{D}_1 = C^\infty(e^{itA_1})$ and $\mathcal{D}_2 = C^\infty(e^{itA_2})$ leads to $\mathcal{D}_1 \bar{\otimes} \mathcal{D}_2 = C^\infty(e^{itA_1} \otimes e^{itA_2})$, since $A_1 \otimes 1 + 1 \otimes A_2$ is the infinitesimal generator of the unitary group $t \longmapsto e^{itA_1} \otimes e^{itA_2}$.
Note that the Fréchet space $\mathcal{D}_1 \bar{\otimes} H_2$ is the natural domain of $\mathcal{B}_1 \otimes \mathbb{C}_{id}$, and thus is exactly the Fréchet space of all σ-convergent families $x = (x_i)_{i\in I}$ with values in $\mathcal{D}_1$, where I is an index set such that $H_2 = L^2_{\mathbb{C}}(I)$ (see [1], for example). In general, one has inclusions

$$\mathcal{D}_1 \hat{\otimes}_\pi \mathcal{D}_2 \subset \mathcal{D}_1 \bar{\otimes} \mathcal{D}_2 \subset \mathcal{D}_1 \hat{\hat{\otimes}}_\varepsilon \mathcal{D}_2 ,$$

where π stands for the projective tensor product and ε for the biequicontinuous topology (note that $\mathcal{D}_1 \hat{\hat{\otimes}}_\varepsilon \mathcal{D}_2$ is the set of compact linear maps from $\mathcal{D}_1'$ into $\mathcal{D}_2$ due to the reflexivity of $\mathcal{D}_1$). Inclusions are reduced to topological equality for

$\mathcal{D}_1$, $\mathcal{D}_2$ nuclear spaces, and $\mathcal{D}_1 \bar{\otimes} \mathcal{D}_2$ is a nuclear space, too.

<u>Lemma 3.6</u>. The Sequence $(A_1)_i \otimes (A_2)_i$ is cofinal in the $*$-algebra $\mathcal{B}_1 \otimes \mathcal{B}_2$ of definition 3.2. Thus,

$$\mathcal{D}_1 \bar{\otimes} \mathcal{D}_2 = \bigcap_{i \geq 0} \left((A_1)_i^{-1} \otimes (A_2)_i^{-1}\right) (H_1 \otimes H_2).$$

If $\mathcal{D}_{01}$ (resp. $\mathcal{D}_{02}$) is a linear set contained, and dense, in the Fréchet space $\mathcal{D}_1$ (resp. $\mathcal{D}_2$), then the algebraic tensor product $\mathcal{D}_{01} \otimes \mathcal{D}_{02}$ is dense in the Fréchet space $\mathcal{D}_1 \bar{\otimes} \mathcal{D}_2$.

The interest in $(A_1)_i \otimes (A_2)_i$ comes from the formula $((A_1)_i \otimes (A_2)_i)(\mathcal{D}_1 \otimes \mathcal{D}_2) = \mathcal{D}_1 \otimes \mathcal{D}_2$ - implying $((A_1)_i \otimes (A_2)_i)\ (\mathcal{D}_1 \bar{\otimes} \mathcal{D}_2) = \mathcal{D}_1 \bar{\otimes} \mathcal{D}_2$. For simplicity, we will confuse $(A_1)_i \otimes (A_2)_i$ and tis closure $\overline{(A_1)_i \otimes (A_2)_i}$.

<u>Proof</u>. A linear set $\mathcal{D}_0$ is dense in the Fréchet space $\mathcal{D}$ (with $\mathcal{D} = \mathcal{D}_1$ or $\mathcal{D}_2$) iff all A_i are essentially self-adjoint on $\mathcal{D}_0$. It is now a known fact that $(A_1)_i \otimes 1 + 1 \otimes (A_2)_i$ is essentially self-adjoint on the algebraic tensor product of two cores in the Hilbert spaces $\mathrm{Dom}(\overline{A_1})_i$ and $\mathrm{Dom}(\overline{A_2})_i$, respectively, thus showing the density of $\mathcal{D}_{01} \otimes \mathcal{D}_{02}$ in $\mathcal{D}_1 \bar{\otimes} \mathcal{D}_2$. The estimations

$$1 \otimes (A_2)_i + (A_1)_i \otimes 1 \leq 2(A_1)_i \otimes (A_2)_i \ ;$$

$$(A_1)_i \otimes (A_2)_i \leq ((A_1)_i \otimes 1 + 1 \otimes (A_2)_i)^2$$

on the set $\mathcal{D}_1 \otimes \mathcal{D}_2$ hold by continuity on the natural domain $\mathcal{D}_1 \bar{\otimes} \mathcal{D}_2$ of $\mathcal{B}_1 \otimes \mathcal{B}_2$, due to the first proof, thus showing that $(A_1)_i \otimes (A_2)_i$ is a cofinal sequence. Since the inverse of the closure of this operator is the Hilbert tensor product $(A_1)_i^{-1} \otimes (A_2)_i^{-1}$, we get the formula relative to $\mathcal{D}_1 \bar{\otimes} \mathcal{D}_2$.

<u>Definition 3.3</u>. Let $\beta_1 \in B(\mathcal{D}_1, \mathcal{D}_2)$ and $\beta_2 \in B(\mathcal{D}_2, \mathcal{D}_2)$. The tensor product $\beta_1 \otimes \beta_2$ is the continuous sesquilinear form on $(\mathcal{D}_1 \bar{\otimes} \mathcal{D}_2) \times (\mathcal{D}_1 \bar{\otimes} \mathcal{D}_2)$ defined by

$$(\beta_1 \otimes \beta_2)(x \otimes y, z \otimes t) = \beta_1(x, z)\, \beta_2(y, t)$$

for all $x, z \in \mathcal{D}_1$, $y, t \in \mathcal{D}_2$. If $\beta_1 = B_1(A_1)_i \times (A_1)_i$, $\beta_2 = B_2(A_2)_j \times (A_2)_j$, for some $i \geq 0$, $j \geq 0$, then

$$\beta_1 \otimes \beta_2 = (B_1 \otimes B_2)\left((A_1)_i \otimes (A_2)_j\right) \times \left((A_1)_i \otimes (A_2)_j\right).$$

The fact that $\beta_1 \otimes \beta_2 \in B(\mathcal{D}_1 \bar{\otimes} \mathcal{D}_2, \mathcal{D}_1 \bar{\otimes} \mathcal{D}_2)$ follows from the density of $\mathcal{D}_1 \otimes \mathcal{D}_2$ into $\mathcal{D}_1 \bar{\otimes} \mathcal{D}_2$.

The finite linear combination of all $\beta_1 \otimes \beta_2$, with β_1 (resp. β_2) moving in A_1 (resp. in A_2), is denoted by $A_1 \otimes A_2$, and $A_1 \otimes A_2$ is easily seen to satisfy condition II - with cofinal sequence $(A_1)_i \otimes (A_2)_i$ - when A_1 and A_2 satisfies this condition. The ultraweak closure of $A_1 \otimes A_2$ relative to $\mathcal{D}_1 \bar{\otimes} \mathcal{D}_2$ will be denoted by $A_1 \bar{\otimes} A_2$. For some details, we refer to proposition 4.5.

Proposition 3.15. Take $\mathcal{D}_1$ and $\mathcal{D}_2$, endowed with their respective Fréchet topology. Then $\mathcal{D}_1 \otimes \mathcal{D}_2$ is a Schwarz (resp. nuclear) space iff $\mathcal{D}_1$ and $\mathcal{D}_2$ are Schwartz (resp. nuclear) spaces.

Proof. We first assume that $\mathcal{D}_1$ and $\mathcal{D}_2$ are Schwartz (resp. nuclear) spaces. We need to show that $(A_1)_j \otimes (A_2)_j$ has a compact inverse in the Hilbert space $H_1 \otimes H_2$, for sufficiently large j. Using proposition 2 of [10] - see also theorem 5.3 of this paper - we may choose $j \in \mathbb{N}$ such that $(A_1)_j^{-1}$ and $(A_2)_j^{-1}$ are compact (resp. nuclear) in H_1 and H_2, respectively. Introducing the spectral decompositions $(A_\alpha)_j^{-1} = \sum_{i=1}^{\infty} \lambda_i^\alpha < \cdot\, , \varphi_i^\alpha > \varphi_i^\alpha$ of these operators (i.e., $\alpha = 1$ or 2 with $\lambda_1^\alpha \geq \cdots \geq \lambda_n^\alpha \geq \cdots > 0$ and $\{\varphi_n^\alpha\,;\ n \geq 1\}$ orthonormal basis in H_α, we get

$$(A_1)_j^{-1} \otimes (A_2)_j^{-1} = \sum_{i,k}^{\infty} \lambda_i^1 \lambda_k^2 < \cdot\, , \varphi_i^1 \otimes \varphi_k^2 > \varphi_i^1 \otimes \varphi_k^2\, ,$$

clearly compact (resp. nuclear) in $H_1 \otimes H_2$. Conversely, let $\mathcal{D}_1 \bar{\otimes} \mathcal{D}_2$ be Schwartz spaces and $j \geq 0$, with $(A_1)_j^{-1} \otimes (A_2)_j^{-1}$

a compact operator in $H_1 \otimes H_2$. Let $y \in H_2$, with $(A_2)_j^{-1} y \neq 0$, and Ω_1 be the unit ball in H_1. Due to the relative compactness of $(A_1)_j^{-1}\Omega_1 \otimes (A_2)_j^{-1} y$ in the Hilbert space $H_1 \otimes H_2$ we get the relative compactness of $(A_1)_j^{-1}\Omega_1$ in H_1, i.e., $(A_1)_j^{-1}$ is compact in H_1 and $\mathcal{D}_1$ is a Schwartz space; and similarly for $(A_2)_j^{-1}$. When $\mathcal{D}_1 \bar{\otimes} \mathcal{D}_2$ is nuclear, $(A_1)_j^{-1} \otimes (A_2)_j^{-1}$ is nuclear in $H_1 \otimes H_2$ for suitable $j \geq 0$, thus we first find that $(A_k)_j^{-1}$, $k = 1, 2$, are compact. In the preceding representation of $(A_1)_j^{-1} \otimes (A_2)_j^{-1}$ we know that $\Sigma_{i,k} \lambda_i^1 \lambda_k^2 < +\infty$, implying that $\Sigma_{i=1}^{\infty} \lambda_i^k < +\infty$, $k = 1, 2$, i.e., $\mathcal{D}_k$ is nuclear.

Remark 3.2. All preceding considerations on tensor products extend directly to a finite product of spaces A_i with natural domain $\mathcal{D}_i$, $1 \leq i \leq n$, and notations $\mathcal{D}_1 \bar{\otimes} \cdots \bar{\otimes} \mathcal{D}_n = \bar{\otimes}_{i=1}^{n} \mathcal{D}_i$ $\quad A_1 \bar{\otimes} \cdots \bar{\otimes} A_n = \bar{\otimes}_{i=1}^{n} A_i$ makes sense. The usual associativity property of these tensor products is a straightforward fact.

Proposition 3.16. Let H_1, H_2 be Hilbert spaces and $H = H_1 \otimes H_2$. Let A be ultraweakly closed with condition II and natural domain $\mathcal{D}$ - dense in H. If $A_{id} = L(H_1) \otimes \mathbb{C}_{H_2}$ one has $A = B(\mathcal{D}_1, \mathcal{D}_1) \bar{\otimes} \mathbb{C}_{H_2}$ and $\mathcal{D} = \mathcal{D}_1 \bar{\otimes} H_2$ for a suitable $\mathcal{D}_1$ in H_1.

Proof. Viewing H as a direct sum of copies of H_1, we see each operator $T \otimes I$ corresponds to a direct sum of copies of T. Using $A_i^{-1} \in L(H_1) \otimes \mathbb{C}_{H_2}$, it follows that $\mathcal{D}$ induces on a given H_1 a dense domain $\mathcal{D}_1$ and an injective operator $0 \leq B_i \leq \mathrm{Id}$, satisfying $A_i^{-1} = B_i \otimes 1$, $B_i\mathcal{D}_1 = \mathcal{D}_1$ and $\mathcal{D}_1 = \bigcap_{n\geq 0} B_i^{-n}(H_1)$. In this optic, we get the proposition by help of density theorems.

An isolated result on manipulation of closed operators and their adjoints is now indicated.

Lemma 3.7. Let H be a Hilbert space and S be a closed (densely defined) linear operator with adjoint S^*. Let Θ be the linear operator in $H \oplus H$, defined by $\mathrm{Dom}\,\Theta \equiv \mathrm{Dom}\,S \oplus \mathrm{Dom}\,S^*$ and $\Theta(x, y) = (S^*y, Sx)$ for $(x, y) \in \mathrm{Dom}\,\Theta$. Then, Θ is a self-adjoint operator.

Using the property that Θ^2 is a self-adjoint operator, and noting that Dom Θ^2 = Dom(S^*S) + Dom(SS^*) and $\Theta^2(x, y) = (S^*Sx, SS^*y)$ for $(x, y) \in$ Dom Θ^2, we thus find a simple proof for the following result (due to von Neumann): S^*S is a (densely defined) self-adjoint operator.

Proof. It is straightforward that Θ is densely defined, so we need to determine the adjoint operator Θ^*. Clearly Dom Θ^* is the set of $(u, v) \in H \oplus H$ such that

$$|\Theta(x, y) \cdot (u, v)| \leq M\|(x, y)\|$$

for all $(x, y) \in$ Dom Θ, where M is a finite constant depending on (u, v). Taking $y = 0$ (resp. $x = 0$), we find that

$$|(Sx, v)| \leq M\|x\| \quad \text{for all } x \in \text{Dom } S$$

$$(\text{resp. } |(S^*y, u)| \leq M\|y\| \text{ for all } y \in \text{Dom } S^*).$$

Thus

$$v \in \text{Dom } S^* \quad (\text{resp. } u \in \text{Dom}(S^*)^* = \text{Dom } S),$$

and

$$(Sx, v) = (x, S^*v) \qquad (\text{resp. } (S^*y, v) = (y, Su))$$

gives Dom $\Theta^* \subset$ Dom $S \oplus$ Dom S^* and $\Theta \subset \Theta^*$ due to

$$\Theta(x, y) \cdot (u, v) = (y, Su) + (x, S^*v) = (x, y) \cdot \Theta(u, v).$$

Conversely, $u \in$ Dom S (resp. $v \in$ Dom S^*) leads to an estimation of the type

$$|(S^*y, u)| \leq M_1\|y\| \quad \text{for all } y \in \text{Dom } S^*$$

$$(\text{resp. } |(Sx, v)| \leq M_2\|x\| \quad \text{for all } x \in \text{Dom } S)$$

for suitable $M_1, M_2 < +\infty$, hence

$$|(Sx, v) + (S^*y, u)| \leq (M_1 + M_2)(\|x\|^2 + \|y\|^2)^{\frac{1}{2}}$$

showing that $(u, v) \in$ Dom (Θ^*). In short,
Dom Θ^* = Dom $S \oplus$ Dom S^* = Dom Θ and Θ is self-adjoint.

Proposition 3.17. Let S be a closed linear operator acting in some Hilbert space and S^* be its adjoint.

1°/ Let $n \geq 0$ be an integer. The operator $(SS^*)^nS$ with domain

$$\text{Dom}(SS^*)^nS = \{x \in \text{Dom } S \,|\, Sx \in \text{Dom } (SS^*)^n\}$$

is closed and $\mathrm{Dom}(SS^*)^n S$ is a core of Dom S. The adjoint of $(SS^*)^n S$ is the closed operator $(S^*S)^n S^*$ with domain $\mathrm{Dom}(S^*S)^n S^* = \{x \in \mathrm{Dom}\, S^* | S^*x \in \mathrm{Dom}(S^*S)^n\}$.

2°/ Let $(\alpha_0, \alpha_1, \ldots, \alpha_n) \in \mathbb{R}^{n+1}$ with $\alpha_n \neq 0$. The operator $\alpha_0 S + \alpha_1 SS^*S + \cdots + \alpha_n S(S^*S)^n$ with domain $\mathrm{Dom}(SS^*)^n S$ is closed and coïncides with $Sp(S^*S) = p(SS^*)S$, where $p(x) = \alpha_0 + \alpha_1 x + \cdots + \alpha_n x^n$ for $x \in \mathbb{R}$. Its adjoint is the closed operator $\alpha_0 S^* + \alpha_1 S^*SS^* + \cdots + \alpha_n S^*(SS^*)^n$

3°/ Let f be a Borel real function and p be the polynomial in 2°/. Operators $Sp(S^*S)f(S^*S)$ and $f(S^*S)Sp(S^*S)$ are closeable - densely defined - with identical closures, i.e.,

$$[Sp(S^*S)f(S^*S)] = [f(SS^*)Sp(S^*S)]$$

and the adjoint

$$[Sp(S^*S)f(S^*S)]^* = [S^*p(SS^*)f(SS^*)].$$

For another Borel real function g, one has

$$\Big[[Sp(S^*S)f(S^*S)] + [Sp(S^*S)g(S^*S)]\Big] = \Big[Sp(S^*S)(f+g)(S^*S)\Big]$$

and

$$\Big[[Sp(S^*S)f(S^*S)]g(S^*S)\Big] = \Big[Sp(S^*S)(fg)(S^*S)\Big]$$

with

$$Sp(S^*S)h(S^*S) = [Sp(S^*S)h(S^*S)]$$

for a bounded Borel real function h.

4°/ Let $f_1, f_2, \ldots, f_n, \ldots$ be a sequence of Borel real functions and $p_1, p_2, \ldots, p_n, \ldots$ be a sequence of real polynomials. The linear set $(\mathrm{Dom}\, S) \cap (\cap_{n\geq 0} \mathrm{Dom}\, [Sp_n(S^*S)f_n(S^*S)])$ is dense in the Hilbert space H and is a core of Dom S.

A variant of 3°/ may be obtained by replacing the quantity $Sp(S^*S)$ by $[Sh(S^*S)]$, where h is any Borel real function, leading, for example to

$$\Big[[Sh(S^*S)]f(S^*S)\Big]^* = \Big[[S^*h(SS^*)]f(SS^*)\Big], \cdots.$$

In 2°/, choosing $\alpha_1 = \alpha_2 = \cdots = \alpha_n = 0$, $\alpha_0 = 1$ and f bounded, we get

$$f(S^*S)S^* \subset S^*f(SS^*) = [S^*f(SS^*)]$$

and

$$(f(S^*S)S^*)^* = Sf(S^*S).$$

Proof. Let Θ be the self-adjoint operator of lemma 1. Obviously, Θ^2 is the closed operator defined by $\Theta^2(x, y) = (S^*Sx, SS^*y)$ and one has $\mathrm{Dom}\,\Theta^2 = \mathrm{Dom}(S^*S) \oplus \mathrm{Dom}(SS^*)$. It follows that the closed operator Θ^{2n} is defined by $\Theta^{2n}(x, y) = ((S^*S)^n x, (SS^*)^n y)$ and admits as domain $\mathrm{Dom}(S^*S)^n \oplus \mathrm{Dom}(SS^*)^n$. Using the formula $\Theta^{2n+1} = \Theta^{2n}\Theta$ between the closed operators involved, and noting that the closed operator Θ^{2n+1} satisfies $\Theta^{2n+1}(x, y) = ((S^*S)^n S^*y, (SS^*)^n Sx)$, with $\mathrm{Dom}\,\Theta^{2n+1} = \mathrm{Dom}(SS^*)^n S \oplus \mathrm{Dom}(S^*S)^n S^*$, we find that each operator $(S^*S)^n S^*$ and $(SS^*)^n S$ is closed, the first being the adjoint of the second due to $\Theta^{2n+1} = (\Theta^{2n+1})^*$, hence 1°/. Following the spectral theory of self-adjoint operators, the operator $\tilde{p}(\Theta) \equiv \alpha_0\Theta + \alpha_1\Theta^3 + \cdots + \alpha_n\Theta^{2n+1}$ is self-adjoint on the domain $\mathrm{Dom}\,\tilde{p}(\Theta) \equiv \mathrm{Dom}\,\Theta^{2n+1}$ and obviously acts in $H \oplus H$ following the formula

$$(\tilde{p}(\Theta))(x, y) = ((\alpha_0 S^* + \cdots + \alpha_n S^*(SS^*)^n)y,$$
$$(\alpha_0 S + \cdots + \alpha_n S(S^*S)^n)x)$$

for $(x, y) \in \mathrm{Dom}\,\Theta^{2n+1} = \mathrm{Dom}\,S(S^*S)^n \oplus \mathrm{Dom}\,S^*(SS^*)^n$. Taking in $\mathrm{Dom}\,\Theta^{2n+1}$ vectors of the form $(x, 0)$ and $(0, y)$ respectively, we get closedness of the operators mentioned, and the self-adjointness of $\tilde{p}(\Theta)$ shows that the first operator has to be the adjoint of the second. Now, as $\tilde{p}(\Theta) = \Theta p(\Theta)^2 = p(\Theta^2)\Theta$ holds for these closed operators (since $\mathrm{Dom}\,p(\Theta^2) = \mathrm{Dom}\,\Theta^{2n}$ implies $\mathrm{Dom}\,\Theta p(\Theta^2) = \mathrm{Dom}\,\Theta^{2n+1}$ and $\mathrm{Dom}\,p(\Theta^2) = \mathrm{Dom}\,\Theta^{2n+1}$) - see also [6], p. 1200 - we deduce that $Sp(S^*S) = p(SS^*)S$ are closed operators also coïnciding with $\alpha_0 S + \cdots + \alpha_n S(S^*S)^n$, hence 2°/.

For the other assertions, we will use von Neumann's techniques of operator algebras [17]. Given a self-adjoint operator T, the von Neumann algebra M generated by the spectral projectors of T is abelian, and hence finite. The operator f(T) is closed and affiliated to M since a bounded operator commuting with spectral projectors of T must commute with f(T) - and Dom f(T) is an essentially dense domain relative to M.

The formulas $[f(T) + g(T)] = (f+g)(T)$, $(fg)(T) = [f(T)g(T)]$ may be viewed as an application of lemma 16.4.2 and theorem XV of [17], where it is found that the product of two closed operators affiliated to M is densely defined with a closure affiliated to M.

For the proof of 3°/, taking $T = \Theta$ and $\tilde{p}(x) = \alpha_0 + \alpha_1 x^3 + \dots + \alpha_n x^{2n+1}$, for $x \in \mathbb{R}$, we find that the operators $f(\Theta^2)\tilde{p}(\Theta)$ and $\tilde{p}(\Theta)f(\Theta^2)$ are densely defined with the same closure. Clearly, $f(\Theta^2)$ is the closed operator defined by $f(\Theta^2)(x, y) = (f(S^*S)x, f(SS^*)y)$, with $\mathrm{Dom}\, f(\Theta^2) = \mathrm{Dom}\, f(S^*S) \oplus \mathrm{Dom}\, f(SS^*)$ and $f(\Theta^2)\tilde{p}(\Theta)$ acts as

$$(f(\Theta^2)\tilde{p}(\Theta))(x, y) = (f(S^*S)S^*p(SS^*)y,\ f(SS^*)Sp(S^*S)x)$$

for

$$(x, y) \in \mathrm{Dom}\, f(\Theta^2)\tilde{p}(\Theta) = \mathrm{Dom}\, f(S^*S)S^*p(SS^*) \oplus \mathrm{Dom}\, f(SS^*)Sp(S^*S).$$

Similarly,

$$(\tilde{p}(\Theta)f(\Theta^2))(x, y) = (S^*p(SS^*)f(SS^*)y,\ Sp(S^*S)f(S^*S)x)$$

for

$$(x, y) \in \mathrm{Dom}\, \tilde{p}(\Theta)f(\Theta^2) = \mathrm{Dom}\, Sp(S^*S)f(S^*S) \oplus \mathrm{Dom}\, S^*p(SS^*)f(SS^*).$$

Noting that the closure in $H \oplus H$ of each of these operators is obtained by regrouping the closures in H, we get, due to $[f(\Theta^2)p(\Theta)] = [\tilde{p}(\Theta)f(\Theta^2)]$, that $[Sp(S^*S)f(S^*S)] = [f(SS^*)Sp(S^*S)]$; and $[Sp(S^*S)f(S^*S)]^* = S^*p(SS^*)f(SS^*)$ - these operators being densely defined as mentioned previously - by the self-adjointness of $[\tilde{p}(\Theta)f(\Theta^2)]$. For another Borel real function g, all the operators written below are densely defined and, using [17], satisfy

$$\Big[[\tilde{p}(\Theta)f(\Theta^2)] + [\tilde{p}(\Theta)g(\Theta^2)]\Big] = [\tilde{p}(\Theta)(f + g)(\Theta^2)]$$

and

$$\Big[[\tilde{p}(\Theta)f(\Theta^2)]g(\Theta^2)\Big] = [\tilde{p}(\Theta)(fg)(\Theta^2)],$$

leading to 3°/.

For every $n \geq 0$, one has

$$[\tilde{p}_n(\Theta)f_n(\Theta^2)](x, y) = ([S^*p_n(SS^*)f_n(SS^*)]y, [Sp_n(S^*S)f_n(S^*S)]x)$$

for

$$(x, y) \in \mathrm{Dom}[\tilde{p}_n(\Theta)f_n(\Theta^2)] = \mathrm{Dom}[Sp_n(S^*S)f_n(S^*S)] \oplus \mathrm{Dom}[S^*p_n(SS^*)f_n(SS^*)]$$

and $D \equiv (\mathrm{Dom}\ \Theta) \cap (\cap_{n\geq 0} \mathrm{Dom}\ [\tilde{p}_n(\Theta)f_n(\Theta^2)])$ is essentially dense relative to the von Neumann algebra generated by the spectral projectors of Θ, implying density in $H \oplus H$. Now, by a well known result, it suffices to show that D is stable under the one-parameter unitary group $t \mapsto e^{it\Theta}$, a straightforward fact, since each closed operator $[\tilde{p}_n(\Theta)f_n(\Theta^2)]$ commutes with $e^{it\Theta}$; hence 4°/.

This paragraph will deal with operations known in von Neumann algebras and is purely technical. We will deal with an ultra-weakly closed space $\mathcal{B} = \cup_{i\geq 0} MA_i \times A_i$ with condition II and natural domain $\mathcal{D}$ (notations of theorem 1.2) and we now introduce an involutive algebra M_0 containing the A_i^{-1} which generates M as a von Neumann algebra. Let $\mathcal{B}_0 = \cup_{i\geq 0} M_0A_i \times A_i$ and $\mathcal{B}'$ be the commutant of $\mathcal{B}_0$ calculated in $B(\mathcal{D}, \mathcal{D})$.
A $*$-algebra A with cofinal sequence A_i and condition I is introduced as an intermediate object contained in B_0. The reader may think at A equal to the $*$-algebra generated by all A_i, A_i^{-1} or at A equal to $\mathcal{B}_0 \cap L(\mathcal{D})$. Note that $M_0 \subset L(H)$, $M \subset L(H)$, $A_{id} \subset L(\mathcal{D})$, $M' \subset L(\mathcal{D})$.

1°/ We first consider <u>the operation of induction</u>. Let E be a projector in M' and K be the Hilbert space K = E(H). We denote by $(\mathcal{B}_0)_E$ (resp. $\mathcal{B}_E$) the set

$$(\mathcal{B}_0)_E = \bigcup_{n\geq 0} M_0(A_nE) \times (A_nE);$$

$$\mathcal{B}_E = \bigcup_{n\geq 0} M(A_nE) \times (A_nE).$$

From the density of $E(\mathcal{D})$ into the Hilbert space K, and the relation $E\mathcal{D} \subset \mathcal{D}$, we see that every $T \in \mathcal{B} \cap L(\mathcal{D})$ satisfies $T(E(\mathcal{D})) \subset E(\mathcal{D})$ and $ET = TE$ on $\mathcal{D}$. It follows that A_E coïncide with the set of operators TE = ET with T moving in A and is seen to be a $*$-algebra with <u>natural</u> domain $E(\mathcal{D})$ (see the proposition below) acting in K.

Taking $T = A_i$ or A_i^{-1}, we find that $A_i^{\varepsilon}(E(\mathcal{D})) = E(\mathcal{D})$ for $\varepsilon = \pm 1$. The topology of $E(\mathcal{D})$ is clearly the topology associated with semi-norms:

$$x \in E(\mathcal{D}) \longmapsto \|A_iEx\| = \|EA_ix\| = \|A_ix\|,$$

and therefore is exactly the topology induced by the Fréchet topology of $\mathcal{D}$ on the linear set $E(\mathcal{D})$.

Putting $\mathcal{D}_E = E(\mathcal{D})$, we see that

$$(\mathcal{B}_0)_E \subset \mathcal{B}_E \subset B(\mathcal{D}_E, \mathcal{D}_E),$$

where the last set consists of all continuous sesquilinear forms on $\mathcal{D}_E \times \mathcal{D}_E$.

Proposition 4.1. $\mathcal{D}_E$ is a closed subspace of the Fréchet space $\mathcal{D}$.

Proof. From $E\mathcal{D} \subset \mathcal{D}$, $(1-E)\mathcal{D} \subset \mathcal{D}$, we find $E(\mathcal{D}) = \{x \in \mathcal{D} | Ex = x\}$. Viewing $(1-E)$ as a continuous linear map from the Fréchet space $\mathcal{D}$ into itself, we find that $\mathrm{Ker}(1-E)$ is closed, thus showing the proposition.

One can sometimes write $A_i E = A_i^E$.

Theorem 4.1. The σ-weak closure of $(\mathcal{B}_0)_E$, relative to $\mathcal{D}_E \hat{\otimes} \mathcal{D}_E$ coïncides with $\mathcal{B}_E$ and the commutant $((\mathcal{B}_0)'_E$ of $(\mathcal{B}_0)_E$ calculated in $B(\mathcal{D}_E, \mathcal{D}_E)$ coïncides $\mathcal{B}'_E$ (the set of all ETE with T in $\mathcal{B}'$).

Proof. By definition, $(\mathcal{B}_0)_E$ is the set of sesquilinear forms $BA_iE \times A_iE$ on the topological product $E(\mathcal{D}) \times E(\mathcal{D})$, with B varying in M_0, and i varying in $\mathbb{N}$. We first show that the von Neumann algebra $ME = EM$, acting in $K = E(H)$, which is generated by the involutive algebra EM_0, coïncides with the Hilbert ultraweak closure P of the set $((\mathcal{B}_0)_E)_{id}$ of bounded elements of $(\mathcal{B}_0)_E$. One has $M_0 \subset \mathcal{B}_0$, hence $EM_0 \subset (\mathcal{B}_0)_E$, thus $EM_0 \subset ((\mathcal{B}_0)_E)_{id}$, therefore $EM \subset P$. In order to show the opposite inclusion, it is sufficient to show that $((\mathcal{B}_0)_E)_{id}$ is included in EM. Let $\beta = BA_iE \times A_iE$, with B in M_0 and i in $\mathbb{N}$, be an element of $((\mathcal{B}_0)_E)_{id}$; we note that β can be viewed as the bounded closeable operator $\beta = A_i^+ EBEA_i$ of $L(K)$. For A in the commutant $(ME)' = EM'E$ for ME, we get, for $x \in E(\mathcal{D})$, $y \in E(\mathcal{D})$,

$$\begin{aligned}(A\beta x, y) &= (\beta x, A^*y) = (BA_iEx, A_iEA^*y)\\ &= (BEA_ix, A^*A_iEy) = ((EA^*E)(BE)A_iEx, A_iEy)\end{aligned}$$

and

$$(\beta Ax, y) = (BA_iEAx, A_iEy) = ((BE)(E\ A\ E)A_iEx, A_iEy),$$

since EAE = A, so that $A\beta = \beta A$ on $E(\mathcal{D})$, implying that $\beta \in (EM'E)' = M_E$; hence $P = M_E$. Now, one clearly has

$$\bigcup_{i\geq 0} (EM_0E)A_iE \times A_iE \subset \bigcup_{i\geq 0} ((\mathcal{B}_0)_E)_{id} A_iE \times A_iE \subset$$

$$\subset \bigcup_{i\geq 0} M_E A_iE \times A_iE = \mathcal{B}_E$$

and theorem 1.2 shows that the σ-weak closure of $(\mathcal{B}_0)_E$ relative to $\mathcal{D}_E \hat{\otimes} \mathcal{D}_E$ is $\mathcal{B}_E$. By theorem 1.2 again, the commutant $((\mathcal{B}_0)_E)' = ((\mathcal{B})_E)'$ is a $\star$-algebra, with domain $E(\mathcal{D})$, acting the Hilbert space K, and one clearly has $((\mathcal{B}_0)_E)_{id} = M_E'$. Now, any operator T, which sends $E(\mathcal{D})$ into itself, can be identified in a natural way to an operator T, which sends $\mathcal{D}$ into itself, by the formula T = ETE. Let $T \in (\mathcal{B}_E)'$; one has, for $A \in M_{E'}$, and $x \in E(\mathcal{D})$, $y \in E(\mathcal{D})$,

$$(ATx, y) = (Ax, T^*y),$$

hence

$$(ATE\,\eta, E\xi) = (AE\eta, T^*E\xi)$$

for $\eta, \xi \in \mathcal{D}$, so that

$$(AETE\,\eta, \xi) = (EA\,\eta, T^*E\,\xi),$$

therefore $ETE \in \mathcal{B}'$, i.e., $ETE \in \mathcal{B}_E'$. Conversely, any operator of the form ETE, with T in $\mathcal{B}'$, is easily seen to belong to $\mathcal{B}_E'$; thus proving our theorem.

<u>Corollary 4.1</u>. Let A be a $\star$-algebra, satisfying condition I, with natural domain $\mathcal{D}$, A^σ be its σ-weak closure relative to $\mathcal{D} \hat{\otimes} \mathcal{D}$, and E be a projector of A'. The σ-weak closure of the $\star$-algebra A_E, with natural domain $\mathcal{D}_E = E(\mathcal{D})$, relative to $\mathcal{D}_E \hat{\otimes} \mathcal{D}_E$, coïncides with $(A^\sigma)_E$. The commutant of the $\star$-algebra A_E coïncides with A_E' (the set of all ETE with T in A').

<u>Corollary 4.2</u>. Let A be an ultraweakly closed $\star$-algebra, with natural domain $\mathcal{D}$ and cofinal central sequence A_i, and E be a projector in A or in A'. Then, A_E and A_E' are ultraweakly closed $\star$-algebras, with natural domain $E(\mathcal{D})$ and admit the EA_iE, $i \in \mathbb{N}$, as a cofinal central sequence.

Proof. We first consider the case where $E \in A'$. As seen previously, A_E consists of the set of continuous sesquilinear forms β of type $\beta = BA_iE \times A_iE$ with B in A_{id} and i in $\mathbb{N}$; it is therefore straightforward that $EA_iE = A_iE$ are cofinal in A_E and central. All assertions follows from theorem 1.2, except that the natural domain of A'_E is $E(\mathcal{D})$; however, this point is contained in proposition 4.1. Let us now assume that E is in A. Since the natural domain of A' is $\mathcal{D}$, E is in the commutant $(A')'$ of A', hence $(A')_E$ and $((A')')_E$ are commuting $\star$-algebras with the required properties; this proves the corollary.

2°/ We now turn to the operation of reduction, which appears to be more delicate, and often not possible. Indeed, let us consider a $\star$-algebra A satisfying condition I, with natural domain $\mathcal{D}$ and σ-weak closure A^σ relative to $\mathcal{D} \hat{\otimes} \mathcal{D}$; if E is a projector of the von Neumann algebra $N = (A^\sigma)_{id}$, and $\beta \in A^\sigma$ the sesquilinear form $(x, y) \in \mathcal{D} \otimes \mathcal{D} \to \beta(Ex, Ey)$ is, in general, not defined, since $E(\mathcal{D})$ is not necessarily contained in $\mathcal{D}$. Having, however, an analogue of theorem 4.1 in view, we see that one has to impose the condition $E(\mathcal{D}) \subset \mathcal{D}$ (in which case we can assume that $E \subset A$, replacing A by the $\star$-algebra A_1 generated by A and E). It follows, therefore, from $(EA_iE)(E\mathcal{D}) \subset E(\mathcal{D})$ and $(EA_i^{-1}E)(E(\mathcal{D})) \subset E(\mathcal{D})$, that $(EA_iE)(E\mathcal{D}) = E(\mathcal{D})$; we put $A_i^E = EA_iE$.

Let M_0, M, M', $\mathcal{B}_0$, $\mathcal{B}$, $\mathcal{B}'$, A and K be as in 1°/, and E be a projector of M such that $E(\mathcal{D}) \subset \mathcal{D}$. We define $(\mathcal{B}_0)_E$ (resp. $\mathcal{B}_E$) to be $(\mathcal{B}_0)_E = \cup_{n\geq 0} M_0(EA_nE) \times (EA_nE)$ (resp. $\mathcal{B}_E = \cup_{n\geq 0} M(EA_nE) \times (EA_nE)$: since $A \subset \mathcal{B}_0$, one has $A_E \subset (\mathcal{B}_0)_E$ and one can check that A_E is an involutive algebra with domain $E(\mathcal{D})$ and cofinal sequence A_i^E, acting in the Hilbert space K, and can be viewed as the set of operators ETE, with T varying in A. The topology of $E(\mathcal{D})$ is defined by the semi-norms

$$x \in E(\mathcal{D}) \to \|EA_iEx\| = \|EA_ix\|,$$

with i moving in $\mathbb{N}$ and is less fine than the topology on $E(\mathcal{D})$ induced by the Fréchet topology of $\mathcal{D}$. Equality holds iff there exists, for every integer j, an integer i and a constant $M < +\infty$, such that $\|A_jEx\| \leq M\|EA_iEx\|$ for all $x \in \mathcal{D}$. This property is clearly satisfied when E commutes with the cofinal sequence A_i, in which case $E(\mathcal{D})$ is a Fréchet space, i.e., $E(\mathcal{D})$ is the natural domain of A_E (same proof as in proposition 4.1). In general, the natural domain $\mathcal{D}_E$ of A_E is $\mathcal{D}_E = \cup_{i\geq 0}$ Dom $\overline{EA_iE}$, and $\mathcal{D}_E \supset E(\mathcal{D})$.

Theorem 4.2. Let E be a projector in M such that $E\mathcal{D} \subset \mathcal{D}$. The set $\cup_{i\geq 0} M_E(A_iE) \times (A_iE) = \cup_{i\geq 0} M(EA_iE) \times (EA_iE)$ coïncides with the σ-weak closure of $(\mathcal{B}_0)_E = \cup_{i\geq 0} M_0(EA_iE) \times (EA_iE)$, relative to $\mathcal{D}_E \hat{\otimes} \mathcal{D}_E$.

Proof. One has $M_0 \subset \mathcal{B}_0$, hence $EM_0E \subset (\mathcal{B}_0)_E$, thus $EM_0E \subset ((\mathcal{B}_0)_E)_{id}$. Since M_0 generates the von Neumann algebra M, EM_0E generate the von Neumann algebra $EME = M_E$, which acts in the Hilbert space $K = E(H)$. Denoting by P the Hilbert ultraweak closure of $((\mathcal{B}_0)_E)_{id}$, we get that $M_E \subset P$. Now let $\beta = BA_iE \times A_iE$ be an element of $((\mathcal{B}_0)_E)_{id}$, with B in EM_0E, and i in $\mathbb{N}$. Considering β as a bounded operator in K, we get, for A in M', and x, y in $E(\mathcal{D})$,

$$(\beta Ax, y) = (BA_iEAx, A_iEy) = (BA_iAEx, A_iEy)$$

$$= (BAA_iEx, A_iEy) = (ABA_iEx, A_iEy)$$

and

$$(A\beta x, y) = (\beta x, A^*y) = (BA_iEx, A_iEA^*y)$$

$$= (BA_iEx, A^*A_iEy) = (ABA_iEx, A_iEy),$$

hence β commutes in $L(K)$ with M_E', i.e., $\beta \in (M_E')' = EME$. Thus, $((\mathcal{B}_0)_E)_{id} \subset EME$, which leads to $P = EME$. By theorem 1.2, the σ-weak closure of $(\mathcal{B}_0)_E$, relative to $\mathcal{D} \hat{\otimes} \mathcal{D}$, is $\cup_{i\geq 0} M_E A_i^E \times A_i^E$, thus proving our theorem.

Proposition 4.2. Take $\mathcal{B}_0$, $\mathcal{B}$, E of theorem 4.2, with $K = E(H) \subset \mathcal{D}$. Then, the natural domain $\mathcal{D}_E$ of $(\mathcal{B}_0)_E$ is the Hilbert space K, and $\mathcal{B}_E$ is the reduced von Neumann algebra M_E ; $(\mathcal{B})_E'$ coïncide with M_E'.

In particular, let us consider the case where the cofinal sequence A_i is abelian. The domain $\mathcal{D}$ is essentially dense by [17], therefore it follows that a sequence E_n of projectors of M exists, satisfying:

1°/ For every n, $E_n(H) \subset \mathcal{D}$.

2°/ The set $\cup_{n\geq 0} E_n(H)$ is dense in the Fréchet space $\mathcal{D}$.

3°/ For every n, $\mathcal{B}_{E_n}$ and $(\mathcal{B}')_{E_n}$ are commuting von Neumann algebras.

Proof. By the closed graph theorem, the operator $A_j E$ for $j \in \mathbb{N}$ is bounded in the Hilbert space H, hence there exists $M < +\infty$ such that

$$\|A_j Ex\| \leq M\|x\|$$

for $x \in H$, implying that

$$\|A_j Ex\| \leq M\|Ex\|,$$

which shows that K is the natural domain of $(\mathcal{B}_0)_E$. For the same reason, $\mathcal{B}_E$ consists of bounded operators, hence $\mathcal{B}_E = (\mathcal{B}_E)_{id} = M_E$ by the proof of theorem 4.2. The proof of the assertion is a simple calculation.
It is worthwhile to indicate the easy

Proposition 4.3. Let A be a $\star$-algebra satisfying condition I, with cofinal central sequence A_i and σ-weak closure A^σ and $\mathcal{B}$ an ultraweakly closed subspace of $B(\mathcal{D}, \mathcal{D})$ containing A. Then $A^\sigma \subset \mathcal{B}$ and projectors of $(A^\sigma)_{id}$ send $\mathcal{D}$ into itself.

This proposition also means that the existence of sufficient projectors E, such that $E(\mathcal{D}) \subset \mathcal{D}$ (in theorem 4.2), is ensured as soon as $\mathcal{B}$ admits a cofinal abelian sequence A_i; indeed, it suffices to apply proposition 4.3 with A equal to the $\star$-algebra generated by all A_i, A_i^{-1}.

3°/ We now consider the operation of ampliation. Let $H_1 = L^2_{\mathbb{C}}(I)$, with I as index set, be a second Hilbert space and $\widetilde{H} = H \otimes H_1$ the Hilbert tensor product viewed as a direct sum of copies of H. The Fréchet space of all σ-convergent families with values in $\mathcal{D}$ [11] is exactly the Fréchet space $\mathcal{D} \bar{\otimes} H_1$ of definition 3.2 and is the natural domain of the $\star$-algebra $A \otimes \mathrm{Id}$.
For $\beta \in B(\mathcal{D}, \mathcal{D})$, the ampliation $\pi(\beta) \in B(\mathcal{D} \bar{\otimes} H_1, \mathcal{D} \bar{\otimes} H_1)$ is defined to be $\beta \otimes \mathrm{Id}$ - a particular case of definition 3.3 - and one sees that

$$\beta(x, y) = \sum_{\alpha\in I} \beta(x_\alpha, y_\alpha)$$

for $x = (x_\alpha)_{\alpha \in I}$, $y = (y_\alpha)_{\alpha \in I}$ in $\mathcal{D} \bar{\otimes} H_1$. When $\beta \in L(\mathcal{D})$, this definition is the usual one, i.e., $\pi(\beta) = \beta \otimes \mathrm{Id} = \bigoplus_{\alpha \in I} \beta$ direct sum of copies of β. Finally, for $\beta = BA_i \times A_i$ with $B \in L(H)$ one has $\pi(\beta) = \pi(B)\pi(A_i) \times \pi(A_i)$, denoting by π the map $T \longmapsto T \otimes \mathrm{Id}$.

Proposition 4.4. Let $\mathcal{B}$ be an ultraweakly closed space of $B(\mathcal{D}, \mathcal{D})$ with cofinal sequence A_i and condition II. The ampliation $\pi(\mathcal{B})$ is an ultraweakly closed space of $B(\mathcal{D} \bar{\otimes} H_1, \mathcal{D} \bar{\otimes} H_1)$ with cofinal sequence $\pi(A_i)$ and condition II; one has $\pi(\mathcal{B}_{id}) = \pi(\mathcal{B})_{id}$.

Proof. By definition, $\pi(\mathcal{B})$ consists of all forms $\tilde{\beta}$ in $B(\mathcal{D} \bar{\otimes} H_1, \mathcal{D} \bar{\otimes} H_1)$ which can be written $\tilde{\beta} = \pi(B)\pi(A_i) \times \pi(A_i)$, with B in $\mathcal{B}_{id}$ and i in $\mathbb{N}$. Thus, $\pi(\mathcal{B})$ is stable under maps $\tilde{\beta} \to \pi(A_i)\tilde{\beta}\pi(A_i)$ and $\tilde{\beta} \to \pi(A_i^{-1})\tilde{\beta}\pi(A_i^{-1})$. We first show that $\pi(\mathcal{B}_{id}) = \pi(\mathcal{B})_{id}$. The inclusion $\pi(\mathcal{B}_{id}) \subset \pi(\mathcal{B})_{id}$ is clear. Conversely, any $\tilde{\beta} = \pi(B)\pi(A_i) \times \pi(A_i)$ in $\pi(\mathcal{B})_{id}$ can be extended to $\tilde{H}$, or equivalently, there exists $M < +\infty$ such that

$$|\tilde{\beta}(x, y)| \leq M\|x\| \, \|y\|$$

for all x, y in $\mathcal{D} \bar{\otimes} H_1$. Restricting this formula to each copy of $\mathcal{D}$ (i.e., $x, y \in \mathcal{D}$), we get that $BA_i \times A_i$ is a bounded operator B_i, hence $\tilde{\beta}$ coïncide on each copy of $\mathcal{D}$ with B_i, which is $\tilde{\beta} \in \pi(\mathcal{B}_{id})$. Hence, $\pi(\mathcal{B}_{id}) = \pi(\mathcal{B})_{id}$ is an involutive algebra, so that $\pi(\mathcal{B})$ satisfy condition II. Now, it is known that $\pi(\mathcal{B}_{id})$ is a von Neumann algebra, leading to proposition 4.4.

4°/ The formation of a tensor product has been defined in definition 3.3, and now we only indicate

Proposition 4.5. Let A_1 and A_2 be spaces with condition II; one has $A_1^\sigma \bar{\otimes} A_2^\sigma = (A_1 \otimes A_2)^\sigma$, where σ refers, respectively, to $\mathcal{D}_1 \hat{\otimes} \mathcal{D}_1$, $\mathcal{D}_2 \hat{\otimes} \mathcal{D}_2$, and $(\mathcal{D}_1 \bar{\otimes} \mathcal{D}_2) \hat{\otimes} (\mathcal{D}_1 \bar{\otimes} \mathcal{D}_2)$. For A_1 and A_2 ultraweakly closed, the von Neumann algebra $(A_1 \bar{\otimes} A_2)_{id}$ is the tensor product of the von Neumann algebras $(A_1)_{id}$ and $(A_2)_{id}$. For a normal representation π_1 (resp. π_2) of ultraweakly closed A_1 (resp. A_2) - see definition 7.1 - $\pi_1 \otimes \pi_2$ is a normal representation of $A_1 \bar{\otimes} A_2$.

CHAPTER 5: GELFAND TRANSFORMATION

Let A be an abelian $\star$-algebra, with cofinal sequence A_i and natural domain $\mathcal{D}$, satisfying condition I. By proposition 6.3, we can assume that the spaces $(A_{A_i}, \| \ \|_{A_i})$, $i \in \mathbb{N}$ are Banach spaces: it follows that A_{id} is an abelian $\mathbb{C}^*$-algebra containing Id, and consists of operators sending $\mathcal{D}$ into itself. Let A'_{id} be the Banach space dual of the $\mathbb{C}^*$-algebra A_{id}, and K be the spectrum (i.e., the set of characters) of A_{id}, which is known to be compact for $\sigma(A'_{id}, A_{id})$. We denote by $C(K)$ the Banach space of real continuous functions on the compact space K, with the usual supremum norm, and by

$$\Phi : T \in A_{id} \to \Phi(T) = \hat{T} \in C(K)$$

the classical Gelfand transforms. From lemma 10.1 of [1], the set X of characters of A can be identified to a subset of K. Finally, for every integer i, we put $K_{A_i} = \{\chi \in K \text{ such that } \chi(A_i^{-1}) = 0\}$.

Lemma 5.1. K_{A_i} is a compact subset of K, with an empty interior. The spectrum X of A is the set $X = K - (\cup_i K_{A_i})$.

Proof. If K_{A_i} contains an open set $\Omega \neq \phi$, we can find a function $g \neq 0$, $g \in C(K)$, with support contained in Ω. Let $p = \Phi^{-1}(g)$; clearly, $A_i^{-1}p = pA_i^{-1} = 0$ in A_{id}, hence $p = 0$ (due to $A_i(\mathcal{D}) = \mathcal{D}$), which is a contradiction. The second assertion is formulated in lemma 10.1 of [1].

Corollary 5.1. Let T be a positive bounded operator in A_{id}, with $\|T\| = 1$. Then, for any $\varepsilon > 0$, there exists a character on χ such that $\chi(T) \geq 1 - \varepsilon$.

Proof. The function $\hat{T}$ being continuous on the compact space K admits a maximum at some point $\chi_0 \in K$. Since X is everywhere dense in K, our assertion follows from the continuity of this function.

The Gelfand transformation, extended to the $*$-algebra $\mathcal{A}$, associates to every operator T of $\mathcal{A}$, the function $\hat{T} = \Phi(T)$ defined on X by $\hat{T}(\chi) = \chi(T)$. This correspondence is an involutive and injective homomorphism, which preserves the natural ρ-norms (see [11]), these properties being independent of any topology $\mathcal{C}$ on X. When X has the weak topology $\sigma(\mathcal{A}', \mathcal{A})$ functions so obtained on X are continuous. However, it seems more natural to endow X with topology induced by the compact space X, in order to emphasize the important rôle played by $\mathcal{A}_{id}$.

We first describe the algebra $\Phi(\mathcal{A})$ of functions on X. We begin to look at the particular case of a cofinal sequence A_i finitely generated (namely, there exists a finite number $A_1, \ldots, A_p$ in this set such that powers $A_1^{i_1} \ldots A_p^{i_p}$, for $i_1, \ldots, i_p \in \mathbb{N}$, are cofinal in $\mathcal{A}^+$; we can assume that p = 1 since the operator $A = A_1 \ldots A_p$ satisfies $A(\mathcal{D}) = \mathcal{D}$ and powers A^n are clearly cofinal in $\mathcal{A}^+$.

Proposition 5.1. Let $A \in \mathcal{A}$, with $A \geq Id$, $A(\mathcal{D}) = \mathcal{D}$ and $\mathcal{A} = \cup_{n\geq 0} \mathcal{A}_{A^n}$. Then, X is an open set dense in K. Elements of $\Phi(\mathcal{A})$ are continuous functions on X, this set being endowed with topology induced by the compact set K. In fact, for every $f \in \Phi(\mathcal{A})$, there exists $g \in \mathcal{C}(K)$ and $n \in \mathbb{N}$ such that $f = g \cdot \hat{A}^n$ on X (by definition, $\hat{A} = \Phi(A)$).

This proposition also means that $\Phi(\mathcal{A})$ is the algebra of functions on X, generated by $\hat{A}$ and its powers, and by the set of bounded uniformly continuous functions on X. The proof are obvious.

Proposition 5.2. Let $\mathcal{A} = \cup_i \mathcal{A}_{A_i}$ be an abelian $*$-algebra satisfying condition I, Φ the Gelfand transform of $\mathcal{A}$, and X endowed with topology induced by K. We put $\hat{A}_i = \Phi(A_i)$, for $i \in \mathbb{N}$. The function $\hat{A}_i$ is continuous on X, and can be uniquely extended to a continuous function (denoted always by $\hat{A}_i$) on the open dense set $K - K_{A_i}$. For every $f \in \Phi(\mathcal{A})$, there exists an integer i, and $g \in \mathcal{C}(K)$ such that $f(x) = g(x)\hat{A}_i(x)$ for all $x \in K - K_{A_i}$.

Clearly, any function of the form $g \cdot \hat{A}_i$, with $g \in \mathcal{C}(K)$ comes from an element T of $\mathcal{A}$, since $\mathcal{A}_{id}$ is a $\mathbb{C}^*$-algebra.

Proof. Let $\mathcal{B}$ be the $*$-algebra, with domain $\mathcal{D}$, generated by $\mathcal{A}_{id}$ and by A_i : the natural domain of $\mathcal{B}$ is, in general, bigger than $\mathcal{D}$. We have $\mathcal{B} \subset \mathcal{A}$, hence $\mathcal{B}_{id} \subset \mathcal{A}_{id}$, and by construction $\mathcal{A}_{id} \subset \mathcal{B}$, implying that $\mathcal{A}_{id} = \mathcal{B}_{id}$. Since the compact set K depends only on $\mathcal{A}_{id}$, the preceding proposition can be applied, thus proving our proposition, since any f of $\Phi(\mathcal{A})$ is contained in a suitable $\mathcal{B}$.

Lemma 5.2. For every basic measure ν on K, one has $\nu(\cup_i K_{A_i}) = 0$. For every $T \in \mathcal{A}$ and $x \in \mathcal{D}$, the function $\hat{T}$ is $\nu_{x,x}$-integrable.

It is convenient to denote by $f \in C(K) \to T_f \in \mathcal{A}_{id}$ the inverse of the Gelfand transformation. We also recall that there exists, for every $x, y \in \mathcal{D}$, a measure $\nu_{x,y}$ on K, such that $\nu_{x,y}(f) = (T_f x, y)$, for $f \in C(K)$.

Proof. The Gelfand transformation Φ extends to the set $L^\infty(K, \nu)$ of ν-measurable, essentially bounded complex functions on K by the formula $(T_f x, y) = \int_K f(\zeta)\, d\nu_{x,y}(\zeta)$, for $x, y \in \mathcal{D}$, and it is known that the image obtained is exactly the Hilbert weak closure $\mathcal{A}''_{id}$ of $\mathcal{A}_{id}$. The set $E = \cup_i K_{A_i}$ is a Borel subset of K, and corresponds to a projector p_E of $\mathcal{A}''_{id}$. One has $\hat{A}_i^{-1} \cdot E = 0$ on K, hence $A_i^{-1}\, p_E = 0$, and $p_E\, \mathcal{D} \subset \mathcal{D}$ leads to $p_E = 0$, which is equivalent to $\nu(E) = 0$. Thus, each K_{A_i} is $\nu_{x,y}$-negligible, with $x, y \in \mathcal{D}$. Let us fix an integer i, and define, for every $n \in \mathbb{N}$,

$$E_n = \left\{ x \in K \ ; \ \hat{A}_i^{-1}(x) \geq \frac{1}{n} \right\},$$

χ_{E_n} the corresponding characteristic function, $p_{E_n} = \Phi(\chi_{E_n})$. The sequence $f_n = \hat{A}_i\, \chi_{E_n}$ consists of essentially bounded functions, and converges simply to the function $\hat{A}_i(1-\chi_{K_{A_i}})$.

By definition, $\Phi(f_n) = A_i p_{E_n}$ and, for $x \in \mathcal{D}$,

$$|\nu_{x,x}(f_n)| = |(A_i p_{E_n} x, x)| \leq \|A_i x\|\ \|x\|.$$

Since f_n is increasing, it follows that sup f_n is an integrable function, i.e., $\hat{A}_i$ is $\nu_{x,x}$-integrable. It is now easily seen that any function of the form $g\hat{A}_i$, with g in $L^\infty(K, \nu)$, is $\nu_{x,x}$-integrable, thus proving lemma 5.2.

For simplicity of notation, we introduce

Definition 5.1. Let Z be a locally compact space, ν a measure on Z, and $\hat{A}_i$ be a sequence of ν-measurable real functions (defined almost everywhere on Z), $\hat{A}_i \geq 1$ such that $1/\hat{A}_i \in L_c^\infty(Z, \nu)$ and $\nu(Z_i) = 0$ for every integer i, where $Z_i = \{x \in Z;\ 1/\hat{A}_i(x) = 0\}$. Introducing $X = Z - \cup_i Z_i$, we define $L_c^\infty((A_i), X, \nu)$ to be the set of functions of the form $g \cdot \hat{A}_i$, with g moving in $L_c^\infty(X, \nu) = L_c^\infty(Z, \nu)$, and i moving in $\mathbb{N}$.

In practice, $L_c^\infty((A_i), X, \nu)$, endowed with the usual addition and multiplication of functions on X (modulo ν-negligible sets in X), is an algebra; for this, it is sufficient to know that, for every integer i, there exists $j \geq i$, and $M < +\infty$ such that $\widehat{A_i^2} \leq M\, \widehat{A_j}$.

Proposition 5.3. Let ν, K, $X = K - \cup_{i\geq 0} K_{A_i}$ be as in lemma 5.2. For every $f \in L_c^\infty((\hat{A}_i), X, \nu)$, the formula

$$(T_f x, y) = \int_X f(\zeta)\, d\nu_{x,\,y}(\zeta),$$

with $x, y \in \mathcal{D}$, defines an element T_f of the ultraweak closure A^σ of A. The correspondence $f \to T_f$ is a ρ-norm isomorphism of the algebra $L_c^\infty((\hat{A}_i), X, \nu)$ onto the algebra A^σ.

Proof. The properties mentioned are known for $f \in L_c^\infty(X, \nu)$. By the proof of lemma 5.2, one has, for $x \in \mathcal{D}$, $\nu_{x,x}(\hat{A}_i) = (A_i x, x)$, i.e., $T_{\hat{A}_i} = A_i$. Moreover, for $x, y \in \mathcal{D}$, and E_n of this lemma, one has $E_n\, \nu_{x,A_i y} = E_n\, \hat{A}_i\, \nu_{x,y}$, due to $\hat{A}_i(x) \leq n$ for x in E_n, hence $\nu_{x,A_i y} = \hat{A}_i\, \nu_{x,y}$. Now, for $g \in L_c^\infty(X, \nu)$ one has $(T_g x, y) = \int_X g(\xi)\, d\nu_{x,y}$, hence

$$(T_g x, A_i y) = \int_X g(\xi)\, d\nu_{x,A_i y},$$

and, on the other hand,

$$(A_i T_g x, y) = \int_X g(\xi)\, \hat{A}_i(\xi)\, d\nu_{x,y},$$

which is our formula, with $f = g \cdot \hat{A}_i$. All the other assertions are straightforward.

In order to describe general abelian, ultraweakly closed $\star$-algebras, we now mention the easy

Lemma 5.3. Let M be an abelian von Neumann algebra, and $A \geq Id$ be a self-adjoint operator affiliated to M. We denote by Z a locally compact space with positive measure ν, and Φ an isometric isomorphism (for algebraic structures) from $L_c^\infty(Z, \nu)$ onto M. Then, the set of elements of Z on which the function $\Phi^{-1}(A^{-1})$ vanishes is negligible, and there exists a ν-measurable function $\hat{A}$ such that $\Phi(g\hat{A}) = A\Phi(g)$, for any $g \in L_c^\infty(Z, \nu)$ such that $g\hat{A} \in L_c^\infty(Z, \nu)$.

Theorem 5.1. Let A be an abelian $\star$-algebra, with cofinal sequence A_i and natural domain $\mathcal{D}$, satisfying condition I, A^σ be its ultraweak closure, and M be the von Neumann algebra A_{id}^σ. There exists a locally compact space Z, a positive measure ν on Z, with support Z, a sequence $\hat{A}_i$ of measurable functions on Z, finite almost everywhere, as introduced in definition 5.1, and a ρ-norm isomorphism $\tilde{\Phi}$ of the involutive algebra $L_c^\infty((\hat{A}_i), X, \nu)$ onto the involutive algebra A^σ, which extends the usual isomorphism Φ from $L_c^\infty(Z, \nu)$ onto M, and which satisfies $\tilde{\Phi}(A_i) = \hat{A}_i$ for $i \in \mathbb{N}$.

One must note that functions $\hat{A}_i$ are not necessarily continuous (compare with proposition 5.2); by suitably choosing Z one can, however, find for every function of $L_c^\infty((\hat{A}_i), X, \nu)$, a continuous function on X which coïncides a.e. with the given function.

Corollary 5.2. Take A, Z, ν of theorem 5.1. There exists a sequence E_n of ν-measurable sets, such that $\nu(X - \cup_n E_n) = 0$, and such that $L_c^\infty((\hat{A}_i), E_n, \nu_{E_n}) = L_c^\infty(E_n, \nu_{E_n})$ for every $n \in \mathbb{N}$, the first space consisting of restriction to E_n functions of $L_c^\infty((\hat{A}_i), E_n, \nu)$, with ν_{E_n} the restriction of the measure ν to E_n.

Proofs. We sketch proofs of theorem 5.1 and corollary 5.2. Take Z, Φ, M as in lemma 5.1, with ν the basic measure on Z. Since $A_i^{-1} \in M$, we construct functions $\hat{A}_i^{-1} = \Phi^{-1}(A_i^{-1}) \in L_c^\infty(Z,\nu)$ and define Z_i as the subset of Z where this function vanishes. Then, $\cup_{i\geq 0} Z_i$ is ν-negligible, and $\hat{A}_i$ of lemma 5.3, equal to $1/\hat{A}_i^{-1}$ is a $\nu_{x,x}$-integrable function (see the proof of lemma 5.2).

Adding the proof of proposition 5.3, we get theorem 5.1. Since $\mathcal{D}$ is essentially dense, there exists an increasing sequence p_n of projectors in M, with $p_n(H) \subset \mathcal{D}$. Taking for E_n measurable sets corresponding via Φ to p_n, we get an isomorphism from the reduced von Neumann algebra M_{p_n} onto $L^\infty_{\mathbb{C}}(E_n, \nu_{E_n})$, which is corollary 5.2.

Proposition 5.4. Let A be a self-adjoint operator, $A \geq Id$, and $\mathcal{A}$ be the $*$-algebra generated by A, A^{-1}, Id, with natural domain $\mathcal{D} = \bigcap_{n\geq 0} \mathrm{Dom}(\bar{A}^n)$. Let S be the spectrum of A. Then, the spectrum of the $*$-algebra $\mathcal{A}$ is homeomorphic to S.

In which case, A corresponds by the Gelfand transformation Φ, to the function $f(x) = x$ on S.
Let B be the $\mathbb{C}^*$-algebra generated in $L(H)$ by the involutive algebra $\mathcal{A}_{id}$, and $\mathcal{A}_1 = \bigcup_{n\geq 0} BA^n \times A^n \subset B(\mathcal{D}, \mathcal{D})$; from paragraph 6, $\mathcal{A}_1$ is an abelian $*$-algebra satisfying condition I with natural domain $\mathcal{D}$ and $(\mathcal{A}_1)_{id} = B$. It easily follows that the spectrum of $\mathcal{A}_1$ coïncides with that of $\mathcal{A}$ and lemma 5.1 shows that $\Phi(\mathcal{A})$ contains continuous functions with compact support in S.

Proof. We may assume that A is unbounded. The involutive algebra generated by Id and A^{-1} coïncides with $\mathcal{A}_{id}$, since $\mathcal{A}$ consists of all operators of the form $\Sigma_{-\infty}^{\infty} a_n A^n$, where (a_n) is any sequence of complex numbers, with $a_n = 0$ for sufficiently large $|n|$. The spectrum of the abelian Banach algebra generated by $\mathcal{A}_{id}$ is topologically isomorphic with the spectrum $K \subset [0, 1]$ of A^{-1} and, characters of $\mathcal{A}$ being continuous characters χ on $\mathcal{A}_{id}$ with $\chi(A^{-1}) \neq 0$, must therefore correspond to points $0 \neq \chi \in K$. Noting that A^{-1} corresponds to the function $f(x) = x$ on K, that the spectrum of A^{-1} is the closure in the real line of reals of type $1/\lambda$ with $\lambda \in S$, we prove our proposition, since $x \longmapsto 1/x$ is a homeomorphic form $]0, 1]$ onto $[1, +\infty[$.

Proposition 5.5. Let A be an (unbounded) self-adjoint operator, with spectrum S, $\mathcal{D} = \bigcap_{n\geq 0} \mathrm{Dom}(\bar{A}^n)$, and $\mathcal{B}$ be a subspace of $B(\mathcal{D}, \mathcal{D})$, stable under involution and containing all powers A^n. Let a_n be a sequence of positive reals. Then, there exists on $\mathcal{B}$ a positive linear form f satisfying $f(A^n) = a_n$ for all n iff there exists a positive measure p on $\mathbb{R}$, with support in S such that $\int x^n\, dp(x) = a_n$ for all n ('moment problem').

Proof. Let $\mathcal{A}$ be the $*$-algebra generated by 1, A, A^{-1}, with natural domain $\mathcal{D}$. Any positive linear form f on $\mathcal{A}$ has a positive extension to $\mathcal{B}$, so that we can assume $\mathcal{A} = \mathcal{B}$. By the preceding proposition, the spectrum of $\mathcal{A}$ coïncides with S. Thus, the existence of p imply the existence of f. Conversely, let be given f a positive linear form such that $f(A^n) = a_n$. If $P = \Sigma_k\, \xi_k\, X^k$ is some polynomial with $P(x) \geq 0$ for $x \in S$, and $\Phi : \mathcal{A} \to C(S)$ denotes the Gelfand transformation, one clearly has $\Phi^{-1}(P) = \Sigma_k\, \xi_k\, A^k \geq 0$ on $\mathcal{D}$, hence $\Sigma_k\, \xi_k\, a_k \geq 0$, which implies the existence of p.

Definition 5.2. Take $(\hat{A}_i)$, Z, ν, as introduced in definition 5.1. The set of ν-measurable functions g on Z, such that $\hat{A}_i g \in L^1_c(Z, \nu)$ for every integer i, is denoted by $L^1_c((\hat{A}_i), Z, \nu)$.

We endow $L^1_c((\hat{A}_i), Z, \nu)$ with the topology given by the sequence of semi-norms

$$f \in L^1_c((\hat{A}_i), Z, \nu) \longmapsto \|\hat{A}_i f\|_1 = \int_Z f\, \hat{A}_i\, d\nu.$$

It is straightforward that $L^1_c((\hat{A}_i), Z, \nu)$ is a Fréchet space under this topology (which is clearly independent of the choice of the cofinal sequence $\hat{A}_i$ of $L^\infty_c((\hat{A}_i), Z, \nu)$).
When no confusion arises, we write A_i instead to $\hat{A}_i$. For ν-measurable real functions f and g, formula $f \leq g$ means $f(x) \leq g(x)$ ν-almost everywhere. Due to the underlying order structure of the function spaces considered, it is convenient to introduce $L^\infty_{\mathbb{R}}((\hat{A}_i), Z, \nu)$, $L^1_{\mathbb{R}}((\hat{A}_i), Z, \nu)$ the real part of spaces previously considered. For simplicity of notation, we will omit the subscript $\mathbb{R}$ or $\mathbb{C}$, and simply write $L^\infty((\hat{A}_i), Z, \nu)$, $L^1((\hat{A}_i), Z, \nu)$.
We endow $L^\infty((A_i, Z, \nu)$ with an obvious ρ-topology defined as follows. For any integer i, $L^\infty_{A_i}$ is the image of $L^\infty(Z, \nu)$

under the map φ_i ; $g \in L^{\infty}(Z, \nu) \mapsto A_i g = \varphi_i(g) \in L^{\infty}((A_i), Z, \nu)$, and has norm $\| \ \|_{A_i}$ in such a way that φ_i becomes an isometric isomorphism. The unit ball of $L^{\infty}_{A_i}$ is the set $[-A_i, A_i]$ of ν-measurable functions g, such that $|g| \leq A_i$. The ρ-topology of $L^{\infty}((A_i), Z, \nu)$ is the locally-convex inductive limit of the sequence of normed space $(L^{\infty}_{A_i}, \| \ \|_{A_i})$. It is not surprising that:

Proposition 5.6. The ρ-topology of $L^{\infty}((A_i), Z, \nu)$ is separated, and any positive linear form on this space is continuous. A fundamental system of ρ-bounded subsets consists of homotheties of intervals $[-A_i, A_i]$, $i \in \mathbb{N}$. The sum

$$L^{\infty}_{\mathbb{C}}((A_i), Z, \nu) = L^{\infty}_{\mathbb{R}}((A_i), Z, \nu) \oplus iL^{\infty}_{\mathbb{R}}((A_i), Z, \nu)$$

is a direct topological sum.

These properties can be shown directly, but they also follow directly from proposition 5.10.

A positive linear form φ on $L^{\infty}((A_i), Z, \nu)$ is called normal iff $\varphi(\sup_{\alpha} h_{\alpha}) = \sup_{\alpha} \varphi(h_{\alpha})$, for any bounded increasing net h_{α} in $L^{\infty}((A_i), Z, \nu)^{+}$. Let $L^{\infty}((A_i), Z, \nu)'$ be the strong dual of $L^{\infty}((A_i), Z, \nu)$. We recall that a fundamental system of semi-norms of this Fréchet space consists of the maps

$$\varphi \in L^{\infty}((A_i), Z, \nu)' \longmapsto \|\varphi\|_{A_i} = \sup_{h \in [-A_i, A_i]} |\varphi(h)|,$$

where i moves in $\mathbb{N}$. Elements in the linear span N of positive normal linear forms on $L^{\infty}((A_i), Z, \nu)$ will be called normal too.

Proposition 5.7. A positive linear form φ on $L^{\infty}((A_i), Z, \nu)$ is normal iff there exists $g \geq 0$ in $L^{1}((A_i), Z, \nu)$ such that $\varphi(h) = \int_Z h\, g\, d\nu$ for all $h \in L^{\infty}((A_i), Z, \nu)$.

Such a function g, when it exists, is clearly unique. We denote by Φ the injective map from $L^{1}((A_i), Z, \nu)$ onto N defined by

$$\Phi(g)h = \int_Z gh\, d\nu$$

for $h \in L^{\infty}((A_i), Z, \nu)$ and $g \in L^{1}((A_i), Z, \nu)$.

Proof. Let $\varphi \geq 0$ normal on $L^{\infty}((A_i), Z, \nu)$. The restriction of φ to $L^{\infty}(Z, \nu)$ is also normal, thus there exists a unique $h \geq 0$, $h \in L^1(Z, \nu)$, such that

$$\varphi(g) = \int_Z gh \, d\nu \qquad (*)$$

for all $g \in L^{\infty}(Z, \nu)$. We choose an integer $i \geq 1$ and show that $A_i h \in L^1(Z, \nu)$. Let, for $n \in \mathbb{N}$,

$$B_n = \{\xi \in Z \mid A_i(\xi) \leq n\}.$$

It is clear that $\nu(Z - \cup_n B_n) = 0$ and that $A_i \chi_{B_n} h$ increases simply ν a.e. to $A_i h$. Due to

$$\varphi(A_i \chi_{B_n}) = \int_Z A_i \chi_{B_n} h \, d\nu$$

and

$$\sup_n \int_Z A_i \chi_{B_n} h \, d\nu \leq \varphi(A_i) < +\infty,$$

it follows that $A_i h$ is ν-integrable. Since this holds for all i, we get $h \in L^1((A_i), Z, \nu)$. Formula (*) clearly holds for all h in $L^{\infty}((A_i), Z, \nu)$. Finally, unicity of h is straightforward, being ensured by the restriction of φ to $L^{\infty}(Z, \nu)$.

Conversely, let $h \geq 0$, $h \in L^1((A_i), Z, \nu)$. From $A_i h \geq 0$ and $A_i h \in L^1(Z, \nu)$, we get that each linear form

$$g \in L^{\infty}(Z, \nu) \longmapsto \int_Z g \, A_i h \, d\nu$$

is normal, and hence prove the normality of the form

$$g \in L^{\infty}((A_i), Z, \nu) \rightarrow \int_Z g \, h \, d\nu,$$

since the correspondence

$$g \in L^{\infty}(Z; \nu) \longmapsto \varphi_i(g) = A_i g \in L^{\infty}((A_i), Z, \nu)$$

exchanges increasing nets and their least upper bounds.

Theorem 5.2. Let N be endowed by the strong dual topology of $L^{\infty}((A_i), Z, \nu)'$. Then, N is a Fréchet space and the correspondence Φ from $L^1((A_i), Z, \nu)$ onto N (of proposition 5.7) is an isomorphism between Fréchet spaces.

Proposition 5.8. Let f be a hermitian linear form in N. Then, there exists $f_1 \geq 0$, $f_2 \geq 0$, $(f_1, f_2) \in N \times N$ such that $f = f_1 - f_2$ and

$$\|f\|_{A_i} = \|f_1\|_{A_i} + \|f_2\|_{A_i} \qquad (**)$$

for all integer i.

Unicity of a couple $(f_1, f_2) \in N \times N$, such that $f = f_1 - f_2$, $f_1 \geq 0$, $f_2 \geq 0$, is ensured as soon as (**) holds for a particular integer i.

Proof. Let $f \in N$, and $g \in L^1((A_i), Z, \nu)$ such that $f = \Phi(g)$. Let μ be the bounded measure $\mu = g \cdot \nu$, and $g^+ = \sup(g, 0)$, $g^- = \sup(-g, 0)$. Introducing $\mu^+ = g^+ \cdot \nu$ and $\mu^- = g^- \cdot \nu$, formula $\mu = \mu^+ - \mu^-$ clearly corresponds to the Jordan decomposition of measure μ, and formula $\|\mu\| = \|\mu^+\| + \|\mu^-\|$ (on the space $L^{\infty}(Z, \nu)$), is relation (**) for i = 0 (since $A_0 = 1$). Now, from $(A_i g)^+ = \sup(A_i g, 0) = A_i g^+$ and a similar formula for $(A_i g)^-$ we get, from $A_i g = A_i g^+ - A_i g^-$, that $A_i g^+ \nu$ and $A_i g^- \nu$ are mutually singular, hence $\|A_i \mu\| = \|A_i \mu^+\| + \|A_i \mu^-\|$ which is formula (*) with $f_1 = \Phi(g^+)$, $f_2 = \Phi(g^-)$.

Let us now assume the existence of a couple $(f_1', f_2') \in N \times N$, such that $f = f_1' - f_2'$ and $\|f\|_{A_k} = \|f_1'\|_{A_k} + \|f_2'\|_{A_k}$ for some integer k. Let φ_k^{-1} be the map $\varphi_k^{-1}(g) = g/A_k$. Putting, for $j = 1, 2$, $\psi_j = f_j \circ \varphi_k^{-1}$ $\psi_j' = f_j' \circ \varphi_k^{-1}$ and $\psi = f \circ \varphi_k^{-1}$, we get, on the space $L^{\infty}(Z, \nu)$

$$\|\psi\| = \|\psi_1\| + \|\psi_2\| = \|\psi_1'\| + \|\psi_2'\|$$

and

$$\psi = \psi_1 - \psi_2 = \psi_1' - \psi_2' ,$$

leading to $\psi_j = \psi'_j$; $j = 1, 2$ on $L^\infty(Z, \nu)$. Thus, f_j and f'_j coïncident on $L^\infty_{A_k}$, and hence on $L^\infty((A_i), Z, \nu)$ by an easy application of Lebesgue's theorem, thus showing the unicity of the decomposition.

Proof of theorem 5.2. Let M be a bounded subset of N. For each f in M, let $f = f_1 - f_2$ be the decomposition of proposition 5.8, and M_1 (resp. M_2) be the set of f_1 (resp. f_2) so obtained. From $\|f_j\|_{A_i} \leq \|f\|_{A_i}$ for any integer i, and $j = 1, 2$, we get that M_1 and M_2 are bounded in N with $M \subset M_1 - M_2$. Thus, the positive cone N^+ of N induces a bounded decomposition in N, therefore N^+ induces an open decomposition [9]. By [9] , N is a Fréchet space as soon as any increasing Cauchy sequence φ_n in N^+ converges. Let $h_n = \Phi(\varphi_n)$ for $n = 0, 1, 2, \ldots$; clearly, $h_n \leq h_{n+1}$ for $n \geq 0$. One has, for $m \geq n$, $\|\varphi_m - \varphi_n\|_{A_i} = \varphi_m(A_i) - \varphi_n(A_i)$, which shows that, for every $i \geq 0$, $h_n A_i$ is an increasing Cauchy sequence in $L^1(Z, \nu)$. If $h \in L^1(Z,\nu)$ denotes the limit function obtained for $i = 0$ it is easily seen that h_n tends to h in the Fréchet space $L^1((A_i), Z, \nu)$ and that $\varphi = \Phi^{-1}(h) \in N$ is the limit of the sequence φ_n. Finally, in order to show that Φ is a topological isomorphism, it is sufficient to show that the bijective map Φ is continuous. But, this point is implicitly contained in the preceding lines.

We now turn to the determination of the dual $L^1((A_i), Z, \nu)$.

Proposition 5.9. Take Z, A_i, ... of definition 5.2, ν being a σ-finite positive measure. Then, for every $h \in L^\infty((A_i), Z, \nu)$ the linear functional $T(f) = \int_Z hf \, d\nu$ is continuous on the Fréchet space $L^1((A_i), Z, \nu)$. Conversely, any continuous linear form T on $L^1((A_i), Z, \nu)$ comes from an element $h \in L^\infty((A_i), Z, \nu)$, via formula $T(f) = \int_Z fh \, d\nu$ (with $f \in L^1((A_i), Z, \nu)$.

The proof given below shows that T moves in an equicontinuous set iff h moves in a bounded subset of $L^\infty((A_i), Z, \nu)$.

Proof. Since functions A_i are ν-measurable on Z, it follows from the Lusin and Egoroff theorems, and the fact that ν is σ-finite, that there exists a ν-negligible set $N \subset Z$, with

$N \supset \cup_i Z_i$ and a partition of Z-N by a sequence of compact sets K_n such that, for any integers i and n, the restriction of the function A_i to K_n is continuous. We denote by $L^1(K_n, \nu)$ the space of ν-integrable functions on K_n, almost everywhere equal to zero on $Z-K_n$. Since each function A_i is ≥ 1 and bounded on each K_n, $L^1(K_n, \nu)$ coïncides with the set of functions of $L^1((A_i), Z, \nu)$ a.e. equal to zero on $Z-K_n$. We also introduce $\Sigma_{n\geq 0} L^1(K_n, \nu)$ the space of functions obtained by taking finite sums of functions, each of them belonging to some $L^1(K_n, \nu)$. For a subset $A \subset Z$, χ_A denotes the corresponding characteristic function.

Let be given an element h in $L^\infty((A_i), Z, \nu)$. There exists an integer i and a constant $C < +\infty$ such that $|h| \leq CA_i$ a.e. on Z. For $f \in L^1((A_i), Z, \nu)$ one has $A_i f \in L^1(Z, \nu)$ hence

$$|T(f)| = |\int_Z hf \, d\nu| \leq C \int_Z |A_i f| \, d\nu = C\|A_i f\|_1$$

shows that T belongs to the dual space of $L^1((A_i), Z, \nu)$.

Conversely, let T be a continuous linear form on $L^1((A_i), Z, \nu)$. One can find, from continuity, an integer i and $M < +\infty$ such that

$$|T(f)| \leq M\|A_i f\|_1$$

for all $f \in L^1((A_i), Z, \nu)$. We will choose $M = 1$ for simplicity. Since $L^1(K_n, \nu)$ is included in $L^1((A_i), Z, \nu)$, we get

$$|T(g)| \leq \|A_i g\|_1$$

for $g \in L^1(K_n, \nu)$. Noting that each function A_i is continuous on the compact set K_n, we can find a constant $m_i(K_n)$ such that $A_i \leq m_i(K_n)$ on K_n.
Therefore, for g in $L^1(K_n, \nu)$,

$$|T(g)| \leq m_i(K_n)\|g\|_1,$$

hence there exists a unique h_n in $L^\infty(K_n, \nu)$ such that

$$T(g) = \int_{K_n} h_n g \, d\nu$$

for all $g \in L^1(K_n, \nu)$ (of course, $\|h_n\|_\infty \leq m_i(K_n)$). Let h be the function defined almost everywhere, equal to h_n on each K_n. We first show that $|h| \leq A_i$ a.e. on each K_n or, equivalently, $|h| \leq A_i$ a.e. on Z. Indeed, if $|h| > A_i$ on some measurable set $X \subset K_n$, with $\nu(X) \neq 0$, there exists $X_1 \subset X$, X_1 compact with $\nu(X_1) \neq 0$ such that h is continuous on X_1 and $|h| > A_i$ on X_1; moreover, X_1 can be chosen such that $h > 0$ (resp. $h < 0$) on X_1. From continuity of the function $|h|/A_i$ on the compact set X_1, there exists $\alpha > 1$ such that $|h| \geq \alpha A_i$ on X_1. One clearly has $\chi_{X_1} \in L^1(K_n, \nu)$ and, assuming (without loss of generality) that $h > 0$, we get simultaneously

$$T(\chi_{X_1}) = \int_{K_n} h \, \chi_{X_1} \, d\nu \geq \alpha \int_{K_n} A_i \, \chi_{X_1} \, d\nu = \alpha \|A_i \, \chi_{X_1}\|_1 > 0$$

and

$$|T(\chi_{X_1})| \leq \|A_i \, \chi_{X_1}\|_1,$$

which is clearly impossible; hence, $|h| \leq A_i$ a.e. on Z.
It remains to show that $T(g) = \int_Z hg \, d\nu$. Let $g \in L^1((A_i), Z, \nu)$; we put $g_n = \Sigma_{i=1}^n g_i \chi_{K_i} \in \Sigma_{n\geq 0} L^1(K_n, \nu)$. Then, $|hg_n| \leq |hg|$ and the sequence hg_n converge simply a.e. to hg. Clearly, $|hg| \leq |A_i g|$ and, $|A_i g|$ being ν-integrable, it follows from the Lebesgue theorem that

$$\int_Z hg_n \, d\nu \to \int_Z hg \, d\nu.$$

In the same manner, from $|A_i(g_n - g)| \leq 2|A_i g|$, we obtain

$$\|A_i(g_n - g)\|_1 = \int_Z A_i(g_n - g) \, d\nu$$

tends to zero, implying that Tg_n tends to Tg. Now, from construction h, formula $Tg_n = \int_Z g_n h \, d\nu$ holds and, taking limits on both sides, we obtain $Tg = \int_Z gh \, d\nu$, thus proving our proposition.

<u>Proposition 5.10</u>. Take Z, ν, A_i of definition 5.1, with $\nu \geq 0$ and $L_c^\infty((A_i), Z, \nu)$ be an algebra for the usual addition and multiplication of functions. Let H be the Hilbert space $L_c^2(Z, \nu)$ and $\mathcal{D}$ be the linear subset of H consisting of g in H such that $A_i g \in H$ for all $i \geq 0$. We denote by $\mathcal{A}$ the set of all T_f, with $T_f g = fg$ for $f \in L_c^\infty((A_i), Z, \nu)$ and $g \in \mathcal{D}$. Then,

1°/ $\mathcal{A}$ is an ultraweakly closed $*$-algebra with natural domain $\mathcal{D}$ (in particular $\mathcal{D}$ is dense in H) such that $\mathcal{A}' = \mathcal{A}$

2°/ the correspondence $f \to T_f$ is an isomorphism from $L_c^\infty((A_i), Z, \nu)$ onto $\mathcal{A}$, which preserves ρ-norms.

<u>Proof</u>. For $f \in L^\infty(Z, \nu)$, let $\dot{T}_f$ be the operator $\dot{T}_f g = fg$, for $g \in H$ and M be the von Neumann algebra for all $\dot{T}_f$, with f in $L^\infty(Z, \nu)$. We first show that $\mathcal{D}$ is dense in H. Each formula $T_{A_i} g = fg$ defines a self-adjoint operator on the domain $\mathcal{D}_i = \{g \in H | A_i g \in H\}$ which is affiliated to M, since spectral projectors of T_{A_i} corresponds to suitable projectors of M. As M is finite, $\mathcal{D}_i$ is an essentially dense domain, as well as $\mathcal{D} = \cap_{i \geq 0} \mathcal{D}_i$ by [17]. It is straightforward that an f in $L^\infty((A_i), Z, \nu)$ leads to a bounded operator T_f in H iff f is in $L^\infty(Z, \nu)$, hence we get $M = \mathcal{A}_{id}$. Since $A_j^2 \in L^\infty((A_i), Z, \nu)$ for every $j \geq 0$, one has $A_j \mathcal{D} \subset \mathcal{D}$, and the product fg being in $\mathcal{D}$ as soon as f is in $\mathcal{D}$ and g in $L^\infty(Z, \nu)$, we find that $A_j^{-1} \mathcal{D} \subset \mathcal{D}$, hence $A_j \mathcal{D} = \mathcal{D}$. We point out that $\mathcal{D}$ is a Fréchet space for the semi-norms $f \to \|A_i f\|_H$ each $T_{A_i} \geq Id$ being a closed operator on $\mathcal{D}_i$. Thus, $\mathcal{A}$ satisfies condition I, being obviously a $*$-algebra on $\mathcal{D}$, and is ultraweakly closed by theorem 1.1. Since $M' = M$, and since $A_i \in \mathcal{A}'$, we find that A_i is cofinal in $\mathcal{A}'$, hence $\mathcal{A}' = \cup_{i \geq 0} M' A_i \times A_i = \mathcal{A}$. The second assertion follows from the definition of respective ρ-topologies.

A discrete version of Gelfand transformation is concerned with

Theorem 5.3. Let A be an abelian $\star$-algebra satisfying condition I_0 or I with natural domain $\mathcal{D}$, and X be the set of characters of A. We assume that $\mathcal{D}$ is a Schwartz space for its topology. Then:

1°/ to each $\varphi \in X$ corresponds a vector $x_\varphi \in \mathcal{D}$ such that $\|x_\varphi\| = 1$ and $Tx_\varphi = \varphi(T)x_\varphi$ for all $T \in A$.

2°/ for $\varphi, \psi \in X$ with $\varphi \neq \psi$, x_φ is orthogonal to x_ψ. In particular, X is at most countable.

3°/ for $\varphi \in X$, the set H_φ of elements x in $\mathcal{D}$, such that $Tx = \varphi(T)x$ for all $T \in A$, is finite dimensional and the Hilbert space H is the direct Hilbert sum of all H_φ, $\varphi \in X$.

For an abelian A satisfying condition I, the reader can easily check that $\mathcal{D}$ is a Schwartz space iff A contains an operator $T \geq \mathrm{Id}$ with a compact inverse.

Proof. It is sufficient to treat the case of a $\star$-algebra A satisfying condition I; indeed, the existence of sufficient characters on such an A induces a similar property on any $\star$-subalgebra. Let A_i be, as usual, a cofinal sequence in A^+, and $\varphi \in X$. Since φ is positive and $\mathcal{D}$ is a Schwartz space, there exists, from proposition 10 of [10], a sequence x_n in $\mathcal{D}$ such that $\varphi = \Sigma_{i=1}^{\infty} \omega_{x_i,x_i}$. It is easily seen that φ is extremal in the convex set of states on A, hence all x_i are equal to zero, except one of them: x_φ. For simplicity, we write $x_\varphi = x$.
One has

$$\varphi(1) = 1 = (x, x)$$

and

$$\varphi(1) = \varphi(A_i^2)\varphi(A_i^{-2}) = (A_i^2 x, x)(A_i^{-2} x, x)$$

hence

$$1 = (x, x) \leq (A_i x, A_i^{-1}x) \leq \|A_i x\| \; \|A_i^{-1}x\| \leq 1$$

implies that $A_i x \neq 0$, $A_i^{-1}x \neq 0$, and $A_i x = \lambda A_i^{-1}x$ with suitable $\lambda \neq 0$; in fact, $A_i^2 x = \lambda x$ and $A_i^2 \geq 0$ leads to $\lambda > 0$. We now find that

$$\varphi(A_i^2) = (A_i^2 x, x) = \lambda(x, x),$$

hence $\lambda = \varphi(A_i^2)$, so that $A_i^2\, x = \varphi(A_i^2)x$ for all i.

Let A be the Banach algebra completion of $(A_{id}, \| \|_{id})$. The restriction of φ to A_{id} is a continuous character on that space, hence extends uniquely as a character (denoted φ) on A; clearly, $\varphi = \omega_{x,x}$ on A. For a hermitian operator T in A, such that $\|T\| < 1$, (1+T) has a bounded inverse in A, thus the preceding calculation gives

$$(1+T)^2 x = \varphi((1+T)^2)x$$

and, due to $(1+T)^{\frac{1}{2}} \in A$,

$$(1+T)x = \varphi((1+T))x,$$

hence, taking T = 0 and substracting,

$$Tx = \varphi(T)x.$$

This formula holds for any T in A, as seen by taking homotheties of T. For an element T in A, we choose an integer i such that $A_i^{-1} \; T \; A_i^{-1} \in A_{id}$, thus

$$\varphi(A_i^{-1} \; TA_i^{-1})x = A_i^{-1} \; TA_i^{-1}x$$

implies that

$$\varphi(A_i^{-2}) \; \varphi(T)x = A_i^{-2} \; Tx;$$

hence, composing each side by A_i^2,

$$\varphi(T) \; \varphi(A_i^{-2}) \; A_i^2 x = Tx$$

and, from $A_i^2 x = \varphi(A_i^2)x$, we finally find that $\varphi(T)x = Tx$. Taking $\psi \neq \varphi$ in X, we obtain, for $T = T^* \in A$,

$$\varphi(T)(x_\varphi, x_\psi) = (Tx_\varphi, x_\psi) = (x_\varphi, Tx_\psi) = \psi(T)(x_\varphi, x_\psi),$$

hence $(x_\varphi, x_\psi) = 0$ in H.

We now show 3°/. Let $\mathcal{B}$ be the σ-weak closure of A and M the von Neumann algebra $\mathcal{B}_{id}$. We know that $\mathcal{B}$ is a ⋆-algebra with natural domain $\mathcal{D}$ that $M\mathcal{D} \subset \mathcal{D}$, $M'\mathcal{D} \subset \mathcal{D}$ and formula BTu = TBu holds for every $B \in M'$, $T \in \mathcal{B}$ and $u \in \mathcal{D}$. It follows that, for $x \in H_\varphi$,

$$TBx_\varphi = BTx_\varphi = B\varphi(T)x_\varphi = \varphi(T)Bx_\varphi,$$

i.e., $B(H_\varphi) \subset H_\varphi$. Thus, the projecteur E onto the closure (in H) of H_φ belongs to $M \subset M'$. Therefore, by proposition 4.1, the natural domain of the $*$-algebra A_E is the closed subspace $E(\mathcal{D})$ of the Fréchet space $\mathcal{D}$. Now, in the Hilbert space $K = E(H)$, the bounded operator $(A_{i+1}E)^{-1}(A_i E)$ coïncides with the homothety of ratio $\varphi(A_{i+1})\varphi(A_i)^{-1}$, and must be a compact operator by proposition 2 of [10], since $E(\mathcal{D})$ is a Schwartz space, implying that K is finite dimensional.

Let H_1 be the Hilbert direct sum of all H_φ, $\varphi \in X$, and $H_2 = H \ominus H_1$ be the orthogonal complement, with p_2 the corresponding projector. For $z \in H_2$, one has $(z, x) = 0$ for all $x \in H_\varphi$, hence $(z, Tx) = 0$ for all $T = T^*$ in A_{id}, thus showing that $T(H_2) \subset H_2$. As before, we find that $p_2 \in M$. We now introduce a character χ on M such that $\chi(p_2) \neq 0$ and $\chi(A_i^{-1}) \neq 0$ for all $i \geq 0$ (proposition 10 [1] applied to M_{p_2}). Such a character extends as a character χ on $\mathcal{B}$, thus there exists $\zeta \in \mathcal{D}$ such that $T\zeta = \chi(T)\zeta$ for all T in $\mathcal{B}$. From $p_2 \in \mathcal{B}$, we get $p_2\zeta \in \mathcal{D}$, and for $T \in \mathcal{B}$, $Tp_2\zeta = \chi(T)p_2\zeta$ gives $p_2\zeta \in H_1$. But, $p_2\zeta = \chi(p_2)\zeta \neq 0$, which shows that $H_1 = H$.

We deduce

Corollary 5.3. Let A be an ultraweakly closed abelian $*$-algebra satisfying condition I, with Schwartz domain $\mathcal{D}$. Then A is $*$-isomorphic to a space $L^\infty((A_i), \mathbb{N}, \nu)$, where ν is positive bounded measure on $\mathbb{N}$.

Remark 5.1. Properties 1°/ and 2°/ of theorem 5.3 holds for an abelian $*$-algebra on which characters are ultraweakly continuous (i.e., $\mathcal{D}$ is not assumed to be a Schwartz space). Theorem 5.3 has a direct relationship with developments in eigenfunctions of regular differential operators. For example, let Ω be a bounded open set in some $\mathbb{R}^n$ and D be a regular elliptic operator with smooth coefficients. By the Garding-Visik theorem, D^k will be essentially self-adjoint on $C_c^\infty(\Omega)$ for $k \geq 1$ and Rellich theorem implies that $\mathcal{D} \equiv \cap_{n\geq 0} \text{Dom } \bar{D}^n$ is a Schwartz space. By proposition 3.5, $C_c^\infty(\Omega)$ is dense in the Fréchet space $\mathcal{D}$ and the Sobolev lemma implies that $\mathcal{D}$ consists of C^∞-functions. Theorem 5.3 shows that eigenvalues of D corresponds to characters of the abelian $*$-algebra A generated by $(1 + D^2)^{\pm 1}$. More generally, given a differential operator D on a C^∞-real manifold V, one may introduce the Sobolev space $W^s(V)$ as the domain of $\overline{(1 + D^+D)}^{s/2}$ and the Fréchet space $W^\infty(V) = \cap_{s\geq 0} W^s(V)$. It is straightforward that $\overline{1 + D^+D}$ is a homeomorphism form $W^\infty(V)$ (resp. the strong dual $W^\infty(V)'$) onto itself, exchanging exactly W^i with some W^j. In practice, $(1 + D^+D)^{-1}$ is compact, i.e., W^∞ is a Schwartz space, and D^k is essentially self-adjoint for all $k \geq 0$ on a natural domain. Especially of interest is the action of a real Lie group G on a C^∞-manifold V with D equal to the Laplacian and $\mathcal{D}$ the space of C^∞-vectors of the unitary representation of G associated to the action of G in $L^2(V, dx)$ where dx is an invariant measure (G is often a C.C.R. group and V a suitable G/Γ).

CHAPTER 6: COFINAL CENTRAL SYSTEMS AND DERIVATIONS

Take an involutive algebra M_0 of bounded operators acting in the Hilbert space H, M_0 containing operators $\bar{A}_n^{-1}$ for $n \in \mathbb{N}$ (with $A_0 = Id$, $A_n\mathcal{D} = \mathcal{D}$ and $\mathcal{D} = \bigcap_{n\geq 0} \bar{A}_n^{-1}(H)$ by hypothesis), such that the center of M_0 contains all $\bar{A}_n^{-1}$. Then it is clear that $M_0 \subset L(\mathcal{D})$, and it follows that $A = \bigcup_{n\geq 0} M_0A_n \times A_n$ is a $\star$-algebra, with natural domain $\mathcal{D}$ satisfying condition I, with all A_i as a cofinal central system.

Proposition 6.1. Let A be a $\star$-algebra admitting a cofinal central subset (A_i), with domain $\mathcal{D}$. Then:

1°/ The multiplication $(S, T) \in (A, \lambda) \times (A, \lambda) \longmapsto ST \in (A, \lambda)$ is bilinear and jointly continuous.

2°/ If $A_i\mathcal{D} = \mathcal{D}$ for every $i \in \mathbb{N}$, the multiplication $(S, T) \in (A, \rho) \times (A, \rho) \to ST \in (A, \rho)$ is jointly continuous, and the involution $T \to T^*$ is λ-continuous on A.

The reader will note that we do not require that A satisfies condition I.

Proof. Given $S \in A$, we denote by $\pi(S)$ (resp. $\delta(S)$) the map $V \in A \to \pi(S)V = SV \in A$ (resp. $V \in A \to \delta(S)V = VS \in A$). In order to simplify notations, we will write $\|V_1\| \leq \|V_2\|$ whenever $\|V_1x\| \leq \|V_2x\|$ for all x in $\mathcal{D}$, with $V_1, V_2 \in A$, and $|V_1| \leq V_2$ whenever $|(V_1x, x)| \leq (V_2x, x)$ for all x in $\mathcal{D}$. Since (A, λ) (resp. (A, ρ)) is a DF-space, it is sufficient to show that $(S, T) \to ST$ is hypocontinuous [7].

Let S be a bounded subset of (A, λ); there exists an integer i and a finite constant M such that $\|T\| \leq M\|A_i\|$ for all T of S. In order to show the equicontinuity of maps $\pi(T)$, $\delta(T)$, for T moving in S, one has to show that restrictions of these maps to some A^{A_k} have this property, for all k in $\mathbb{N}$. But $\|R\| \leq \|A_k\|$ implies that $\|RT\| \leq \|A_kT\|$ and, since A_k belongs to the center of A, $\|A_kT\| = \|TA_k\| \leq M\|A_iA_k\|$, therefore maps $\delta(T)$ for T varying in S are equicontinuous, the injection of $A^{A_iA_k}$ into (A, λ) being continuous. Now, for

$T \in S$ and R in A such that $\|R\| \leq \|A_k\|$, one has

$$\|\pi(T)R\| = \|TR\| \leq M\|A_i R\| = M\|RA_i\| \leq M\|A_h A_i\|$$

which is equicontinuity of $\pi(T)$, for T varying in S; this shows 1°/ to be true.

For the second assertion, we note that the operators A_i^2 are cofinal in A^+, and obviously in the center of A. Let S be a bounded subset in (A, ρ); there exists $i \in \mathbb{N}$ and $M < +\infty$ such that $|T| \leq MA_i^2$ for all T in S. The positive cone of $A_{A_i^2}$ induces an open decomposition in this order-unit ordered vector space, so that one can assume that S consists of positive operators. For $T \in S$, we find that $(T\varphi, \varphi) \geq 0$ for $\varphi \in \mathcal{D}$, hence $(T(\varphi + \lambda\psi), \varphi + \lambda\psi \geq 0$ for $\varphi, \psi \in \mathcal{D}$ and $\lambda \in \mathbb{C}$, so that

$$|\mathrm{Re}(T\varphi, \psi)|^2 \leq (T\varphi, \varphi)(T\psi, \psi)$$

(Re denotes the real part), and, taking $i\psi$ in place of ψ,

$$|(T\varphi, \psi)|^2 \leq 2(T\varphi, \varphi)(T\psi, \psi).$$

We now show that, for $T \in S$, restrictions of maps $\pi(T)$ to $A_{A_k^2}$ are equicontinuous. We first see that $|R| \leq A_k^2$ with $R = R^* \in A$ implies that $A_h^{-1} R A_h^{-1}$ is bounded with norm ≤ 1, hence $\|(A_h^{-1} R A_h^{-1})^2\| \leq 1$, leading to $|R^2| \leq A_h^4$ since A_h^{-1} commutes with elements of A. Now, $|R| \leq A_h^2$ and $0 \leq T \leq A_i^2$ imply, for $\varphi, \psi \in \mathcal{D}$, that

$$|(TR\varphi, \psi)|^2 \leq 2(TR\varphi, R\varphi)(T\psi, \psi)$$

and, taking $\varphi = \psi$,

$$|(\pi(T)R\varphi, \varphi)|^2 = |(TR\varphi, \varphi)|^2 \leq 2(A_i^2 R\varphi, R\varphi)(A_i^2\varphi, \varphi)$$

$$\leq 2(R^2 A_i\varphi, A_i\varphi)(A_i^2\varphi, \varphi) \leq 2(A_i A_h^4 A_i\varphi, \varphi)(A_i^2\varphi, \varphi)$$

$$\leq 2(A_i^2 A_h^4\varphi, \varphi)^2,$$

which is equicontinuity of the maps $\pi(T)$ from $A_{A_h^2}$ into $A_{A_i^2 A_h^4}$.

Since $T \to T^*$ is continuous from (A, ρ) into itself, it follows equicontinuity of $\delta(T)$, $T \in S$, from (A, ρ) into itself, hence hypocontinuity of the multiplication, implying continuity by [7]. Finally, let $T \in A$ such that $\|T\| \leq \|A_i\|$.

From $TA_i = A_iT$ and $T^*A_i = A_iT^*$ on $\mathcal{D}$, it follows that $A_i^{-1}T = TA_i^{-1}$ and $A_i^{-1}T^* = T^*A_i^{-1}$ on $\mathcal{D}$. Clearly $\|\bar{T}x\| \leq \|\bar{A}_ix\|$ for all $x \in \mathrm{Dom}(\bar{A}_i) \subset \mathrm{Dom}(\bar{T})$. From $\bar{A}_i(\mathrm{Dom}\ \bar{A}_i) = H$, we get that $\bar{T}\bar{A}_i^{-1}$ is the continuous extension to the Hilbert space H of the bounded operator $B = TA_i^{-1}$. It is immediate that $T^*A_i^{-1}$ coïncide on $\mathcal{D}$ with the adjoint B^* of B, hence $\|T^*x\| \leq \|A_ix\|$ for all x in $\mathcal{D}$, which is continuity of the involution in λ-topology; thus proving our proposition.

Lemma 6.1. Let A be a $*$-algebra with natural domain $\mathcal{D}$ on which topologies λ and ρ coïncide. For a $*$-algebra $\mathcal{B}$ with natural domain $\mathcal{D}$ with $\mathcal{B} \subset A$, topological spaces $(\mathcal{B}, \rho)$ and $(\mathcal{B}, \lambda)$ are identical.

Proof. Let A_i be a cofinal sequence in $\mathcal{B}$; clearly, A_i is cofinal in A. By theorem 1.2 [11], we need to show that linear forms $\omega_{x,\ y}$, with x in $\mathcal{D}$ and y in H, are ρ-continuous on $\mathcal{B}$. Being continuous on $(A, \lambda) = (A, \rho)$, their restrictions to normed spaces (A_{A_i}, ρ_{A_i}) are continuous. But $(\mathcal{B}_{A_i}, \rho_{A_i})$ is isometric to a subspace of (A_{A_i}, ρ_{A_i}), hence follows continuity and the lemma.

Proposition 6.2. Let $\mathcal{B} = \cup_{i\geq 0} \mathcal{B}^{A_i}$ be a $*$-algebra in which normed spaces $(\mathcal{B}^{A_i}, \lambda_{A_i})$ are Banach spaces. Then operations $(S, T) \in A \times A \rightarrow ST \in A$ and $T \in A \rightarrow T^* \in A$ are continuous, when A has λ-topology.

Proof. We will keep the notation of the last proposition. First, for R, A, T in A, the relation $\|R\| \leq \|A\|$ implies $\|RT\| \leq \|AT\|$, hence $R \in A \rightarrow RT \in A$ is λ-continuous, for T in A. Let us prove that $R \in A \rightarrow TR \in A$ is λ-continuous or, equivalently, that restriction of this map to some A^A is continuous for λ_A norm. Since (A, λ) is ultrabornological, the closed graph theorem may be applied from (A^A, λ_A) into (A, λ). Let R_n be a sequence tending to zero in A^A, such that TR_n tends to some $Y \in A$ for λ-topology. If $\varphi \in \mathcal{D}$, the set of linear forms $\omega_{\varphi,y}$, for y moving in the unit ball H_1 of the Hilbert space H, is an equicontinuous set, and topology λ being the topology on A of uniform convergence on equicontinuous subsets of the dual $(A, \lambda)'$ of (A, λ), it follows that $TR_n\varphi$ tends to $Y\varphi$ in the Hilbert space H. Now, conditions

$R_n\varphi \to 0$ and $TR_n\varphi \to Y\varphi$ in H lead to $Y\varphi = 0$, since T is closeable.
Now, $Y\varphi = 0$ for all $\varphi \in \mathcal{D}$ gives $Y = 0$, which is continuity of $R \to TR$. Finally, the map $(S, T) \in A \times A \to ST \in A$, being separately continuous on a barrelled DF-space, is continuous. The continuity of $T \to T^*$ is obtained in a similar way; indeed, conditions $T_n \to 0$ in some (A^A, λ_A) and $T_n^* \to Y$ in $(\mathcal{B}, \lambda)$ helped with the relation $(T_n\varphi, \psi) = (\varphi, T_n^*\psi)$ for φ, ψ in $\mathcal{D}$, must lead to $0 = (\psi, Y\varphi)$, hence $Y = 0$, which proves our proposition.

Theorem 6.1. Let A be a $*$-algebra satisfying condition I with natural domain $\mathcal{D}$, A' be its commutant with natural domain D, and A^σ be the σ-weak closure of A relatively to $\mathcal{D} \hat{\otimes} \mathcal{D}$. The following properties are equivalent:

1°/ A^σ is a $*$-algebra with natural domain $\mathcal{D}$.

2°/ For every $t \in A^\sigma$, there exists a closeable operator T, such that $\mathrm{Dom}(T) \supset \mathcal{D}$ which satisfies $t(x, y) = (Tx, y)$ for $x, y \in \mathcal{D}$.

3°/ Topologies λ and ρ coïncide on A.

4°/ $(A^\sigma)_{id}(\mathcal{D}) \subset \mathcal{D}$.

Proof. We denote by A_i a cofinal sequence in A^+, such that $A_i^{-1} \in A$, and let M be the von Neumann algebra generated by A_{id}. We know, from [1], that $A^\sigma = \cup_{i\geq 0} MA_i \times A_i$, hence 1°/ <=> 4°/. By definition, A^σ is the set of t in $B(\mathcal{D}, \mathcal{D})$ such that $t(Bx, y) = t(x, B^*y)$ for $x, y \in \mathcal{D}$ and $B \in M'$. Hence, $t \in A^\sigma$ implies that $t(A_i \cdot , A_i \cdot) \in A^\sigma$. Now, $t(x, y) = (Tx, y)$ gives $t(A_ix, A_iy) = (TA_ix, A_iy)$ and from $t(A_i \cdot , A_i \cdot) \in A^\sigma$, we find a closeable operator S with $\mathrm{Dom}(S) \supset \mathcal{D}$ such that, for $x, y \in \mathcal{D}$,

$$(TA_ix, A_iy) = (Sx, y)$$

Hence, $|(Tx, A_iy)| = |(SA_i^{-1}x, y)| \leq c^{te}\|y\|$ leads to $Tx \in \mathrm{Dom}(\bar{A}_i)$, implying that $Tx \in \mathcal{D}$, i.e., $T\mathcal{D} \subset \mathcal{D}$. Taking t in M, we get 4°/. We note that 1°/ $\Rightarrow$ 2°/ is obvious, hence 2°/<=>4°/<=>1°/.
Now, 1°/ =>3°/. Indeed, A^σ being σ-weakly closed in $B(\mathcal{D}, \mathcal{D})$, intervals $[-A_i, A_i]$ are σ-weakly compact, which implies that $(A^\sigma_{A_i}, \rho_{A_i})$ are Banach spaces, hence $\lambda = \rho$ on A^σ by theorem 1.1 [11]. By lemma 6.1, $\lambda = \rho$ on A. We show 3°/ =>1°/. Let

$t \in A^{\sigma}$, with $t = t^*$. By construction of A^{σ}, there exists an integer i and a bounded operator $B = B^* \in M$ such that $t = BA_i \times A_i$. Since M is the von Neumann algebra generated by A_{id}, one can find, by the Kaplansky density theorem, a net B_{α} of hermitian elements in A_{id}, such that $\|B_{\alpha}\| \leq \|B\|$, converging for weak Hilbert topology to B. Put $t_{\alpha} = B_{\alpha}A_i \times A_i$; since $B_{\alpha}\mathcal{D} \subset \mathcal{D}$ one has $t_{\alpha} = A_iB_{\alpha}A_i$ and $t_{\alpha} \in A$. It is clear that t_{α} converges for weak topology (relative to $\mathcal{D} \hat{\otimes} \mathcal{D}$) to t, and from $|t_{\alpha}| \leq \|B\| A_i^2$ and the Ascoli theorem, we find that t_{α} tends to t σ-weakly (relatively to $\mathcal{D} \hat{\otimes} \mathcal{D}$). Let us now fix x in $\mathcal{D}$. Linear forms $h_{x,\alpha} : y \in \mathcal{D} \to t_{\alpha}(x, y)$ are continuous on the prehilbert space $\mathcal{D}$, with norm $\|h_{x,\alpha}\| = \|A_iB_{\alpha}A_ix\|$. From $\lim_{\alpha} t_{\alpha}(x, y) = t(x, y)$, we find that the linear form $h_x : y \in \mathcal{D} \to t(x, y)$ is the simple limit of linear forms $h_{x,\alpha}$. Since λ coïncides with ρ on A, the set of all t_{α}^2 is bounded in (A, ρ); indeed, $(S, T) \in A \times A \to ST \in A$ being continuous sends bounded subsets of $A \times A$ into bounded subsets of A. Thus, there exists $j \in N$ and $M < +\infty$, such that

$$0 \leq t^2 = (A_iB_{\alpha}A_i)^2 \leq M A_j,$$

hence

$$\|A_iB_{\alpha}A_ix\| \leq ((A_iB_{\alpha}A_i)^2x, x)^{\frac{1}{2}} \leq M^{\frac{1}{2}}(A_jx, x)^{\frac{1}{2}}$$

which means that the set of all $h_{x,\alpha}$ is equicontinuous on the prehilbert space $\mathcal{D}$. Therefore, by the Ascoli theorem, the quantity $\lim t_{\alpha}(x, y)$ exists for every y in H. Finally, there exists a unique w in H, such that

$$(BA_ix, A_iy) = t(x, y) = (w, y).$$

The formula $Tx = w$ defines a linear operator T, with $\mathrm{Dom}(T) = \mathcal{D}$ and, from $t(x, x) \in \mathbb{R}$ for x in $\mathcal{D}$, it follows that T is closeable. Since any t in A^{σ} expresses as $t = \frac{t+t^*}{2} + i\,\frac{t-t^*}{2i}$, we associate operators T_1 and T_2 defined for t_1 and t_2, respectively, and it is easily seen that T_1 and T_2 are adjoint w.r.t. each other on $\mathcal{D}$, hence 2°/ is proved.

Corollary 6.1. Let A be a $*$-algebra with natural domain $\mathcal{D}$, and cofinal central sequence A_i such that $A_i\mathcal{D} = \mathcal{D}$, and $\tilde{A}$ be the $\star$-algebra generated by A and all A_i^{-1}. Then, $\widetilde{A}$, A^σ, $\widetilde{A}^\sigma$ are $\star$-algebras with natural domain $\mathcal{D}$, and topologies λ and ρ agree on A, A^σ, $\widetilde{A}$, $\widetilde{A}^\sigma$, the σ-referring to σ-weak closure relatively to $\mathcal{D} \hat{\otimes} \mathcal{D}$. The natural domain D of A' coïncides with $\mathcal{D}$.

In general, for a $\star$-algebra A, it may happen that $\lambda = \rho$ on A, and $\lambda \neq \rho$ on $\widetilde{A}$. However, from theorem 6.1, the equality of λ and ρ on $\widetilde{A}$ implies that $\lambda = \rho$ on $\widetilde{A}^\sigma$.

Proof. Since elements of A send $\mathcal{D}$ into itself, we get, from $\overline{A_i \mathcal{D}} = \mathcal{D}$, that elements of A commute with the bounded operators A_i^{-1} , hence A_i^{-1} is in the center of $\widetilde{A}$. Therefore, A_i^{-1} belongs to the center of $\widetilde{A}_{id}$. Thus, A_i^{-1} is in the center of the von Neumann algebra M generated by $\widetilde{A}_{id}$, which implies that $M\mathcal{D} \subset \mathcal{D}$. Consequently, by theorem 6.1 and by $(\widetilde{A}^\sigma)_{id} = M$, we see that $\widetilde{A}^\sigma$ is a $\star$-algebra contained in $L(\mathcal{D})$, thus $\mathcal{D}$ is the natural domain of $\widetilde{A}$, A^σ, $\widetilde{A}^\sigma$, since $A^\sigma \subset \widetilde{A}^\sigma$. The fact that A^σ is a $\star$-algebra follows from the obvious σ-weak continuity of $T \in A \rightarrow T^* \in A$ and $T \in A \rightarrow ST \in A$, where $S \in A$. Finally, since A_i are in the center of $\widetilde{A}$, one has $A_i \in (\widetilde{A})' = A'$, therefore the sequence A_i is cofinal in A', giving $D = \mathcal{D}$.

Remark 6.1. Let A be any $\star$-algebra with natural domain $\mathcal{D}$ and let B_i be a sequence in A^+ (not necessarily cofinal), $B_i \geq Id$, $B_0 = Id$, $B_i\mathcal{D} = \mathcal{D}$ of elements commuting each other on $\mathcal{D}$. Let

$$\mathcal{B}_0 = \bigcap_{i \in \mathbb{N}} \left\{ S \in A \text{ such that } SB_i = B_iS \text{ and } S^*B_i = B_iS^* \text{ on } \mathcal{D} \right\}$$

and $\mathcal{B}$ be the subset of $\mathcal{B}_0$ consisting of operators smaller than some homothetic of some C_n, where $C_n = B_1^n \ldots B_n^n$, for $n \in \mathbb{N}$. Then it is easily seen that $\mathcal{B}$ is a $\star$-algebra with domain $\mathcal{D}$, in which elements C_n are cofinal and central. However, the natural domain $\mathcal{D}_1$ of $\mathcal{B}$ contains $\mathcal{D}$ and due tu the commutativity of the sequence A_i and the relations $B_i\mathcal{D} = \mathcal{D}$, $C_i\mathcal{D} = \mathcal{D}$, we find that $\bar{B}_i\mathcal{D}_1 = \mathcal{D}_1$, $\bar{C}_i\mathcal{D}_1 = \mathcal{D}_1$, meaning that $\mathcal{B}$ satisfies the corollary. Clearly, B_i^{-1} is in $\mathcal{B}$ as soon as it belongs to A, in which case the σ_1-closure $\mathcal{B}^{\sigma 1}$ of $\mathcal{B}$

(here σ_1 refers to $\mathcal{D}_1 \hat{\otimes} \mathcal{D}_1$) is a $*$-algebra with natural domain $\mathcal{D}_1$ which is contained in the σ-weak closure A^σ of A. Finally, symmetric elements of $\mathcal{B}^{\sigma_1}$ are essentially self-adjoint on $\mathcal{D}_1$ and on $\mathcal{D}$ (compare with [16, 18].

For bounded operators, the construction of the $\mathbb{C}^*$-algebra and von Neumann algebra associated with a given involutive algebra of bounded operators is considered as an automatic process. In this respect, we state

Lemma 6.2. Let A be a $*$-algebra satisfying condition I, with natural domain $\mathcal{D}$ and cofinal sequence A_n. Let M be an involutive subalgebra of the Banach algebra $L(H)$, where H is the Hilbert space completion of $\mathcal{D}$. We define $\mathcal{B} = \cup_{T\in A}$ MT $\times$ T. Then:

1°/ $\mathcal{B}$ is stable under involution, and under the maps $\beta \to \beta A_n \times A_n : \beta \to \beta A_n^{-1} \times A_n^{-1}$: in particular $\mathcal{B} = \cup_{n\geq 0} \mathcal{B}_{id} A_n \times A_n$

2°/ Let M be a $\mathbb{C}^*$-algebra containing A_{id}. Then $\mathcal{B}_{id}$ = M iff the following property is satisfied: given B $\geq$ 0, B $\in$ M and i $\in \mathbb{N}$, such that BA_i is a bounded operator, one has $BA_i \in$ M.

When M is the $\mathbb{C}^*$-algebra generated by A_{id}, it is sufficient in order to get $\mathcal{B}_{id}$ = M, to show that, given B $\geq$ 0 in M with BA_i bounded, there exists for any $\varepsilon > 0$, an element $B' \in A_{id}$ such that $\|Bx - B'x\| \leq \varepsilon \|A_i^{-1}x\|$ for all $x \in \mathcal{D}$.

Proof. Let $\beta \in \mathcal{B}$, and B $\in$ M, T $\in A$, such that β = BT $\times$ T. One has $\beta A_n \times A_n = BTA_n \times TA_n$ and $\beta A_n^{-1} \times A_n^{-1} = BTA_n^{-1} \times TA_n^{-1}$. Since $TA_n \in A$ and $TA_n^{-1} \in A$, we deduce 1°/.
It is now obvious that $M \subset A_{id}\, M\, A_{id} \subset \mathcal{B}_{id}$. Let us assume that the property indicated in 2°/ is satisfied. Let $\beta \in A_{id}$; since $\mathcal{B}_{id}$ contains Id and $\mathcal{B}_{id} = \mathcal{B}_{id}^*$ we may, by replacing β by $\alpha\beta + \lambda$ Id with α, λ suitable complex numbers $\neq 0$, assume that $\beta = \beta^*$ and Id $\leq \beta \leq 3$ Id. Let B $\in$ M and T $\in A$, such that β = BT $\times$ T. We choose an integer n such that $\|Tx\| \leq C\|A_nx\|$ for all $x \in \mathcal{D}$, with $C < +\infty$. Then
$\beta = BTA_n^{-1}A_n \times TA_n^{-1}A_n = B'A_n \times A_n$ with
$B' = (TA_n^{-1})^*B(TA_n^{-1}) \in A_{id}\, M\, A_{id} \subset M$; clearly, $A_n\mathcal{D} = \mathcal{D}$ and $\beta \geq 0$ give $B' \geq 0$. Thus, for x in $\mathcal{D}$, one has

$$(x, x) \leq ((B')^{\frac{1}{2}}A_n x, (B')^{\frac{1}{2}}A_n x) \leq 3(x, x),$$

hence $(B')^{\frac{1}{2}}A_n$ is a bounded operator, therefore $(B')^{\frac{1}{2}}A_n \in M$. However β agrees with the bounded operator $((B')^{\frac{1}{2}}A_n)^*((B')^{\frac{1}{2}}A_n)$, hence $\beta \in M$, showing that $\mathcal{B}_{id} \subset M$ or, equivalently, $\mathcal{B}_{id} = M$. Conversely, let $B \geq 0$, $B \in M$ and $i \in \mathbb{N}$ in such a way that $\beta = BA_i$ is bounded.

Then $A_i^{-1}B \in M$ and $\beta = A_i^{-1}BA_i \times A_i \in \mathcal{B}$. Clearly, β extends continuously to $H \times H$, i.e., $\beta \in \mathcal{B}_{id} = M$, hence $BA_i \in M$.

<u>Proposition 6.3</u>. Let M be the $\mathbb{C}^*$-algebra generated by A_{id}. If A_i is central in A, then $\mathcal{B}_{id} = M$.

<u>Proof</u>. Let $B \in M$ such that $\beta = BA_i \times A_i \in \mathcal{B}_{id}$. One has $\beta + \lambda\, \mathrm{Id} \in \mathcal{B}_{id}$, so that we can assume $\beta \geq 0$. Let B_n be a sequence of hermitian elements in A_{id} converging to B, in the norm of M. From $B\mathcal{D} \subset \mathcal{D}$, we get $\beta = BA_i^2$ on $\mathcal{D}$. One has $(B_n A_i^2)A_i^{-2} = B_n$, $(BA_i^2)A_i^{-2} = B$. Each operator $B_n A_i^2$ is symmetric, hence closeable, with closure $\overline{A_i \quad B_n}$. By [12] p. 207, the sequence $\bar{A}_i^2 B_n$ tends to $\bar{A}_i^2 B$ is the generalized sense, which shows $A_i^2 B_n$ is bounded for sufficiently large n. But $A_i^2 B_n \in A_{id}$, hence $A_i^2 B = \beta \in M$, which is the proposition.

<u>Proposition 6.4</u>. Let A be ultraweakly closed with condition II (and quasi-normable domain) and cofinal central sequence A_i, and T_n be a sequence in A tending to $T \in A$ for ρ-topology. If there exists $a > 0$ and $j \geq 0$ such that $0 < aA_j^{-1} \leq T_n$, then T_n^{-1}, T^{-1} exist and T_n^{-1} tends to T^{-1} in (A, ρ) as $n \to \infty$.

If T_n is a sequence in A ρ-converging to $T \in A$ and T is invertible, it may happen that, for all $n \in \mathbb{N}$, T_n is not invertible, even if $T = \mathrm{Id}$. It follows that the ρ-topology of A, restricted to A_{id}, is, in general, distinct from the norm topology of A_{id}. Moreover, positivity of T does not imply that of T_n, for n sufficiently large. However, (for a quasi-normable domain), given a sequence $T_n = T_n^*$ converging to T in (A, ρ), there exists a decomposition $T_n = T_n^+ - T_n^-$, with $T_n^+ \geq 0$, $T_n^- \geq 0$ for $n \geq 0$, such that T_n^+ (resp. T_n^-)

ρ-converges to T^+ (respectively T^-) in (A, ρ), with $T = T^+ - T^-$; when T is known to be invertible, T_n^+ and T_n^- may be chosen to be invertible. Finally, using proposition 2.3, we see that for every projector E in A commuting with the cofinal sequence, with $E(H) \subset \mathcal{D}_i$, $T_n E_\alpha$ in ρ_i-converging to TE_α, so that the invertibility of $T \geq 0$ implies that of TE_α, showing that $T_n E$ is invertible and positive for sufficiently large n in the Hilbert space E(H); the existence of such a sequence E_α, $\alpha \geq 0$ increasing to identity is known (notation of remark 1.1).

Remark 6.2. Let $A = \cup_{i\geq 0} A_{A_i}$ be ultraweakly closed with condition II and A_i not necessarily central. Any $\beta = \beta^* \in A$ always belongs to some order-unit space A_{A_i}, and thus admits a decomposition $\beta = \beta_1 - \beta_2$ with $\beta_1 \geq 0$, $\beta_2 \geq 0$, $\beta_1, \beta_2 \in A$. Writing $\beta = BA_i^{1/2} \times A_i^{1/2}$ with $B \in A_{id}$, and noting that $A_i^{1/2} \in A$, we get, from the decomposition $B = B^+ - B^-$ with $|B| = B^+ - B^-$. $B^+B^- = B^-B^+ = 0$, a decomposition $\beta = \beta^+ - \beta^-$ with $\|\beta\|_{A_i} = \| |\beta| \|_{A_i}$, where $|\beta| = \beta^+ + \beta^-$, which seems to be the best decomposition for β. When A_i is central, it is obvious that $\beta^+\beta^- = \beta^-\beta = 0$.

Proof. Replacing T_n by $T_n A_j = A_j T_n$, we are reduced to the case where $0 < a \leq T_n$. The set of T_n and T being a bounded set in (A, ρ), there exists $i > 0$ such that $T_n \in A_{A_i}$ and $T \in A_{A_i}$, thus we are reduced to the case where $A_i = \cup_{n\geq 0} A_{id} A_i^n \times A_i^n$, with natural domain $\mathcal{D}_i = \cap_{n\geq 0} \bar{A}_i^{-n}(H)$, is equal to A (noting also that the ρ_i-topology of A_i is finer than that induced by (A, ρ)), and we can simply write $A_i = \Delta$. Each T_n is essentially self-adjoint on $\mathcal{D} \equiv \mathcal{D}_i$, hence $\bar{T}_n^{-1}$ exists and $T_n^{-1}\mathcal{D} = \mathcal{D}$, similarly for T, since $0 < a \leq T$. From $T_n^{-1} - T^{-1} = T_n^{-1}(T-T_n)T^{-1}$, we get, for $x \in \mathcal{D}$,

$$|(T_n^{-1} - T^{-1}x, x)| \leq |((T - T_n)T^{-1}x, T_n^{-1}x)|$$

$$\leq \|(T - T_n)T^{-1}x\| \, \|T_n^{-1}x\|.$$

Since $T_n \to T$ for ρ-topology and $\lambda = \rho$ on A by corollary 6.1 we find $k \geq 0$ and a sequence $\varepsilon(n) \to 0$ such that

$$\|(T_n - T)u\| \leq \varepsilon(n) \|\Delta^k u\|$$

for all $u \in \mathcal{D}$, and $n \geq 0$, thus, as $\|T_n^{-1}\| \leq 1$, we deduce $|T_n^{-1} - T^{-1}| \leq \varepsilon_1(n) \Delta^\ell$ for suitable $\ell \geq 0$ and $\varepsilon_1(n) \to 0$, showing the proposition.

Concerning inverses:

Proposition 6.5. Let $1 \leq T \in A$. If $B \in A$ satisfies $|B-T| \leq T$, then B^{-1} exists and $B^{-1} \in A_{id}$. If $B \in A$ satisfies $|B-T^{-1}| < T^{-1}$, then B^{-1} exists and $B^{-1} \in A$.

Proof. There exists $S \geq Id$, $S \in A$ such that $S^2 = T$. Formula $|B-T| < T$ becomes $|S^{-1} BS^{-1} - Id| < Id$ and, $C = S^{-1} BS^{-1}$ being symmetric, must be invertible in the Banach algebra $L(H)$, with inverse $C^{-1} \in A_{id}$, and $C^{-1}\mathcal{D} = \mathcal{D}$. From $B = CS \times S = S\, C\, S$, we see that $S^{-1}\, C^{-1}\, S^{-1} \in A_{id}$ agrees with the inverse of B. When B satisfies $|B-T^{-1}| < T^{-1}$, we find that $C = T^{\frac{1}{2}}\, B\, T^{\frac{1}{2}} \in A$ is such that $|C-Id| < Id$, and the proof is similar.

Remark 6.3. Our methods appear to be closed to those introduced in [14] concerning an implicit function theorem in Fréchet spaces. Indeed, our proofs are often reduced to an A of the form $A = \cup_n A_{\Delta^n}$, $\Delta \geq Id$, and, putting $\|v\|_r = \|\Delta^r v\|$, we easily see that $T_N \equiv \int_1^N \lambda\, dp_\lambda$ (where $\bar{\Delta} = \int_1^\infty \lambda\, dp_\lambda$) are smoothing operators in sense of [14] p. 1825, since

$$\|T_N v\|_{r+s} \leq N^s \|v\|_r \qquad \|R_N v\|_r \leq \frac{1}{N^s} \|v\|_{r+s}$$

for $N \geq 1$, $r, s \geq 0$, with $R_N = Id - T_N$: also note that $T_N(H)$ consists of differentiable vectors for the one-parameter group $t \to e^{i\Delta t}$. This regularisation process is often used in our paper.

APPENDIX: DERIVATIONS

Let A be a $\star$-algebra, satisfying condition I_0, with natural domain $\mathcal{D}$ and central cofinal sequence A_i. Let δ be a derivation of A, i.e., a linear map δ from A into A such that $\delta(xy) = x \cdot \delta y + \delta x \cdot y$, for $x, y \in A$. Since δ can be extended to a derivation of the $\star$-algebra generated by A and all A_i^{-1}, we can assume that A satisfies condition I: in which case A_{id} is an involutive algebra of bounded operators sending $\mathcal{D}$ into itself, and the restriction of δ to A_{id} is a derivation from A_{id} into A.

The well-known map D defined by Df = f' (f' being the function derivative of the function f) shows that there exist derivations D of abelian $\star$-algebras $A = \cup_i A_{A_i}$, with $A_i^{-1} \in A$, which are not zero; concerning D, one has to identify the set of functions f with the set of operators T_f, defined by $T_f g = fg$ for g chosen in a suitable domain $\mathcal{D}$. For example, one can choose A to be the $\star$-algebra generated by the operators $T_{1/f}$ and $T_{f'}$, with $f(x) = x$ for $x > 1$, acting in the Hilbert space $H = L^2(]1, +\infty[, dx)$, dx being the Lebesgue measure, with $\mathcal{D}$ containing the set of continuous functions with compact support in $]1, +\infty[$.

Taking in account the results relative to derivations in $\mathbb{C}^*$-algebras, it is logical to impose conditions of a topological nature in order to avoid the preceding situation Df = f'.

We denote by A the $\mathbb{C}^*$-algebra completion in $L(H)$ of the involutive algebra A_{id}; clearly, elements of A send the natural domain $\mathcal{D}$ into itself. Let $\hat{A}$ be the $\star$-algebra generated by A and all A_i, A_i^{-1}; it is easily seen that any derivation δ, continuous from (A, ρ) into itself, has a unique continuous extension from $(\hat{A}, \rho)$ into itself, which is a derivation of the $\star$-algebra $\hat{A}$. Thus, one can assume, for the study of continuous derivations, the equality $\hat{A} = A$, or, equivalently, that A_{id} is a $\mathbb{C}^*$-algebra.

<u>Proposition 6.6</u>. Let $A = \cup_i A_{A_i}$ be an abelian $\star$-algebra, with $A_i^{-1} \in A$ for all $i \in \mathbb{N}$, A_{id} being a $\mathbb{C}^*$-algebra. Then, every derivation $\delta : A \to A$ is identically zero.

The proof is an easy adaptation of that stated in [20], relative to abelian $\mathbb{C}^*$-algebras. One has to note the necessity of the assumption $A_i^{-1} \in A$, since Df = f' defines a ρ-continuous derivation of the abelian $*$-algebra A_1, acting in $L^2(]1, +\infty[, dx)$ generated by T_f only, with the same domain $\mathcal{D}$.

Theorem 6.2. Let A be a ultraweakly closed $*$-algebra, satisfying condition II with cofinal central sequence A_i. Then, every derivation δ from A into A is continuous and inner.

As seen in preceding paragraphs, A is ultraweakly closed iff A_{id} is a von Neumann algebra.

Corollary 6.2. Let A be a $*$-algebra (not necessarily commutative) satisfying condition I_0, and A' be the commutant of A. Then every derivation δ from A' into A' is ρ-continuous and inner.

This point is obvious, since A' satisfies the hypothesis of theorem 6.2.

Lemma 6.3. Let A be a $*$-algebra satisfying condition I with cofinal central sequence A_i, and $\delta : A \to A$ be a derivation. Then δ is continuous for ρ-topology iff the restriction of δ to A_{id} is continuous from $(A_{id}, \| \ \|_{id})$ into (A, ρ).

Proposition 6.7. Let A be a $*$-algebra, as in lemma 6.3, satisfying the strict condition of Mackey convergence (see [10 , 7]) such that A_{id} is a $\mathbb{C}^*$-algebra. Then every derivation $\delta : A \to A$ is ρ-continuous.

Theorem 6.3. Let A be a $*$-algebra as in lemma 6.3, with natural domain $\mathcal{D}$, and $\mathcal{B}$ be its ultraweak closure. Any continuous derivation $\delta : A \to A$ can be extended to an ultraweakly continuous derivation $\bar{\delta}$ from $\mathcal{B}$ into $\mathcal{B}$. There exists $T_0 \in \mathcal{B}$ such that $\delta(T) = [T, T_0]$, for $T \in A$.

Combining proposition 6.7 and theorem 6.3, we see that the above mentioned derivation $Df = f'$ is not ρ-continuous when defined on the $*$-algebra generated by T_x and $T_{1/x}$.

Proof of proposition 6.6. Let δ be a derivation on A. Any element of A is a linear combination of positive ones. Now, for $T \geq 0$ in A, there exists j such that $A_j^{-1} T A_j^{-1}$ belongs to A_{id}; and since A_{id} is a $\mathbb{C}^*$-algebra there exists $B \in A_{id}$, $B \geq 0$ such that $B^2 = A_j^{-1} T A_j^{-1}$. Clearly $B A_j^{-1} = A_j^{-1} B$, so that the element $B A_j^{-1}$ satisfies $B A_j^{-1} \geq 0$, $B A_j^{-1} \in A$, and $(B A_j^{-1})^2 = T$.

Let χ be the spectrum of A and $t_0 \in K$. The Gelfand transform realizes A as a suitable algebra of functions on χ. The same arguments as those developed in [20] p. 153, which rest on the existence of a sqare root for any $T \in A$, show that $\delta = 0$.

Proof of lemma 6.3. Let δ be a derivation on A, whose restriction to A_{id} is continuous for the norm $\| \|_{id}$. Denoting by Z the center of A, we get $\delta(Z) \subset Z$. Since $A_i \in Z$, Z is countably dominated, and satisfies condition I. One has $Z_{A_i} = Z \cap A_{A_i}$, which shows that the restriction of δ to (Z, ρ) is ρ-continuous. By proposition 6.6 and proposition 6.3, we find that $\delta(Z) = 0$. It follows that $\delta(A_i^{-1} T A_i^{-1}) = A_i^{-1} \delta(T) A_i^{-1}$, for $T \in A$, hence the restriction of δ to each normed space $(A_{A_i^2}, \| \|_{A_i^2})$ is continuous, i.e., δ is continuous on (A, ρ).

Proof of theorem 6.2. Let Z be the center of A, and A' be the commutant of A. The natural domain D of A' coïncides with $\mathcal{D}$, from corollary 6.1. One has $A_i \in Z$, $A_i^{-1} \in Z$, $\delta(Z) \subset Z$, showing that $Z = \cup_i Z_{A_i}$ satisfies condition I. In particular, Z_{id} coïncides with the center C of the von Neumann algebra A_{id}. Indeed, any element $T \in C$ commutes with A_i^{-1}, and hence with A_i (since $A_i(\mathcal{D}) = \mathcal{D}$), thus commutes with the whole A, showing that $C \subset Z$, i.e., $C \subset Z_{id}$. The converse inclusion is obvious. Thus, Z is ultraweakly closed relatively to $\mathcal{D} \hat{\otimes} \mathcal{D}$ and by proposition 6.6, $\delta(Z) = 0$; it follows that, for $A \in Z$ and $T \in A$, $\delta(AT) = A\delta(T) = \delta(T)A$.

The von Neumann algebra generated by all A_i^{-1} being finite, there exists a sequence E_n of projectors belonging to Z_{id}, such that $E_n(H) \subset \mathcal{D}$, and $(\cup_n E_n(H))^{\perp} = \{0\}$, H being the Hilbert space completion of $\mathcal{D}$. Let E be one of the E_n's. The set A_E of all $TE = ET$, for T moving in A, is a von Neumann algebra (theorem 4.1), and $\delta_E : T \in A_E \to \delta_E(T) = \delta(T) \in A_E$ is clearly a derivation of A_E; thus δ_E is continuous for natural norm of this von Neumann algebra. We also note that $T \in A \to T E \in A_E$ is continuous on (A, ρ).

These points having been stated, if δ is not continuous there exists, by the closed graph theorem, a sequence $S_1, \ldots, S_n, \ldots$ in A tending to zero for (A, ρ), with images

$\delta(S_1)$, ..., $\delta(S_n)$, ... tending to some $A \neq 0$, $A \in A$, for (A, ρ). Thus, we see that, for any projector E of the sequence (E_n), ES_1, ..., ES_n ... tends to zero in A_E, with images $\delta_E(S_1)$, ..., $\delta_E(S_n)$, ... tending to AE in the von Neumann algebra A_E. Due to the continuity of δ_E, we get $AE = 0$, hence $(Ax, x) = 0$ for all $x \in \cup_n E_n(H)$. Since A can be viewed as an element of $B(\mathcal{D}, \mathcal{D})$, and since $\cup_n E_n(H)$ is dense in the Fréchet space $\mathcal{D}$ (proposition 3.10), we find that $A = 0$, thus showing the continuity of δ.

Now, there exists an integer j such that $\delta(A_{id}) \subset A_{A_j^2}$, δ being continuous from the normed spaces $(A_{id}, \| \ \|_{id})$ into the normed space $(A_{A_j^2}, \| \ \|_{A_j^2})$. We define $\lambda(T) = A_j^{-1}\delta(T)A_j^{-1}$, for $T \in A$. By direct calculation, we find that λ is a derivation of A, such that $\lambda(A_{id}) \subset A_{id}$, λ being continuous from the normed space $(A_{id}, \| \ \|_{id})$ into itself. By the nature of derivations in von Neumann algebras, there exists $T_0 \in A_{id}$ such that $\lambda(T) = [T, T_0]$ for $T \in A_{id}$. It follows that $\delta(T) = T(T_0A_j^2) - (T_0A_j^2)T$ for $T \in A_{id}$. For S in A, we choose an integer i such that $A_i^{-1}\ S\ A_i^{-1}$ is bounded and we get:

$$\delta(A_i^{-1}\ S\ A_i^{-1}) = A_i^{-1}\ S\ A_i^{-1}\ T_0\ A_j^2 - T_0\ A_j^2\ A_i^{-1}\ S\ A_i^{-1}\ ,$$

implying that

$$\delta(S) = S(T_0\ A_j^2) - (T_0\ A_j^2)S\ ,$$

hence δ is inner.

Proof of proposition 6.7. Let Z be the center of A. As in the proof of theorem 6.2, one has $\delta(Z) \subset Z$, $A_i \in Z$, $A_i^{-1} \in Z$, and Z_{id} is a $\mathbb{C}^*$-algebra, hence $\delta(Z) = 0$. We can assume that δ is hermitian, the involution in (A, ρ) being continuous. If δ is not continuous, we can find a sequence S_1, ..., S_n, ... tending to zero in (A, ρ), with images $\delta(S_1)$, ..., (S_n), ... tending to $A^1 \neq 0$ in (A, ρ). For any E in Z, one clearly has $\delta(ET) = E\delta(T)$, for $T \in A$. Since (A, ρ) satisfies the Mackey condition of convergence for sequences ([10] p. 765), one can choose an integer j such that operators $X_n = A_j^{-1}\ S_n\ A_j^{-1}$, $\delta(X_n) = A_j^{-1}\ \delta(S_n)A_j^{-1}$ are all bounded for

$n \in \mathbb{N}$ with, moreover, X_n (resp. $\delta(X_n)$) tending to zero (resp. to $A_j^{-1} A^1 A_j^{-1} = A$) in the $\mathbb{C}^*$-algebra A_{id}. As in the proof of [20] p. 154, we are lead to the case where 1 belongs to the spectrum of A: replacing X_n by $Y_n = X_n + 3\|X_n\|.1$ and introducing $H \in A_{id}$ such that $H A H > \frac{1}{2} H^2$, we get $\delta(Y_n) = \delta(X_n)$, and, for every $n \in \mathbb{N}$,

$$H Y_n H \leq 4\|X_n\|^2 H$$

with

$$H Y_n H \leq 8\|X_n\| H A H.$$

Since $\delta(H)$ is not necessarily bounded, one has to choose an integer k such that $\delta(A_k^{-2} H)$ has this property. Of course, any homothetic λH, with $\lambda > 0$, has the properties mentioned above; we take a precise value of λ by requiring that

$$\|A_k^{-1} H^2 A_k^{-1}\| = 1.$$

Thus,

$$A_k^{-1} H Y_n H A_k^{-1} \leq 4\|X_n\| A_k^{-1} H^2 A_k^{-1} \leq 8\|X_n\| A_k^{-1} H A H A_k^{-1}$$

and $\|X_n\|.1 + X_n \geq 0$ gives

$$\frac{A_k^{-1} H Y_n H A_k^{-1}}{4\|X_n\|} \geq \frac{1}{2} A_k^{-1} H^2 A_k^{-1},$$

thus

$$\left\|\frac{A_k^{-2} H Y_n H}{4\|X_n\|}\right\| > \frac{1}{2}$$

Expliciting the quantity $\delta(A_k^{-2} H Y_n H)$, we see that there exists n_0 such that

$$\left\|\delta(A_k^{-2} H Y_{n_0} H) - A_k^{-2} H A H\right\| < \frac{1}{8}$$

Now let B be the $*$-algebra, with domain $\mathcal{D}$, generated by 1, $H Y_{n_0} H$ and all A_k, A_k^{-1}. By corollary 5.1, there exists a character χ on B such that

$$\chi\left(\frac{A_k^{-2}\ H\ Y_{n_0}\ H}{4\|X_{n_0}\|}\right) > \frac{1}{3}$$

and $\chi(1) = 1$. By proposition 1.5 [11] there exists a positive linear form φ on A, which extends χ. Using the Gelfand transform of B, we deduce that $\varphi(\delta B) = 0$ [4] , hence

$$\varphi(A_k^{-1}\ H\ A\ H) < \frac{1}{8}\ .$$

Finally,

$$\varphi(A_k^{-1}\ H\ A\ H) \geq \frac{1}{2}\varphi\left(A_k^{-2}\ \frac{H\ Y_{n_0}\ H}{4\|X_{n_0}\|}\right) \geq \frac{1}{2}\cdot\frac{1}{3} = \frac{1}{6}\ ,$$

which is impossible.

Proof of theorem 6.3. We can assume that δ is hermitian. Let $x = (x_i)$ be a σ-convergent sequence. We need to show that the linear form

$$T \rightarrow \sum_{i=1}^{\infty} (\delta(T)x_i,\ x_i)$$

is σ-weakly continuous or, amounting to the same thing, that its restriction to bounded subsets B of (A, ρ) is weakly continuous. It is easily seen that the maps $T \rightarrow A_j^{-1}\ T\ A_j^{-1}$ and $T \rightarrow A_j\ T\ A_j$ are σ-weakly continuous and the inverse of each other, so that one has to treat the case where $B = [-Id, Id]_A$. Since $\delta(B)$ is a bounded subset of (A, ρ), $\delta(B)$ is a bounded set in some normed space $(A_{A_j^2}, \|\ \|_{A_i^2})$. The map defind by $\lambda(T) = A_i^{-1}\delta(T)A_i^{-1}$, for $T \in A$, is a ρ-continuous derivation of A, with $\lambda(A_{id}) \subset A_{id}$. But A_i induces a topological isomorphism of the set of σ-convergent sequences, so we are lead to the case when $\lambda = \delta$. Finally, by a classical estimation, we get, for $x, y \in \mathcal{D}$, $T \in [-Id, Id]$,

$$|(\lambda(T)x,\ y)| \leq M[(Tx,\ x) + (Ty,\ y)]^{\frac{1}{2}},$$

which is the σ-weak continuity we are looking for. Our theorem follows from theorem 6.2.

CHAPTER 7: NOTION OF REPRESENTATION: THE G.N.S. CASE

We begin with

Theorem 7.1. Let $\mathcal{B} = \cup_i \mathcal{B}_{A_i}$ be an ultraweakly closed subspace of $B(\mathcal{D}, \mathcal{D})$, satisfying condition II, and M be the von Neumann algebra $M = \mathcal{B}_{id}$, acting in the Hilbert space H (completion of $\mathcal{D}$). Let N be a von Neumann algebra acting in some Hilbert space K (in general $K \neq H$), and Φ be a normal homomorphism from M into N such that $\Phi(1) = 1$. We define

$\tilde{\Phi}(A_i) = \overline{\Phi(a_i^{-1})}^{-1}$, and $\mathcal{D}' = \cap_{i\geq 0} \mathrm{Dom}(\tilde{\Phi}(A_i))$. Then:

1°/ $\mathcal{D}'$ is dense in the Hilbert space K, invariant under linear operators $\tilde{\Phi}(A_i)$ and $\Phi(A_i^{-1})$, and depends only on the triplet $(\mathcal{D}, M, \Phi)$.

2°/ The map Φ admits a unique extension as a positive normal map from $\mathcal{B} = \cup_{i\geq 0} M A_i \times A_i$ into $\cup_{i\geq 0} N \tilde{\Phi}(A_i) \times \tilde{\Phi}(A_i)$ and has the following properties

$\tilde{\Phi}(A_j \ \beta \ A_j) = \tilde{\Phi}(A_j)\tilde{\Phi}(\beta)\tilde{\Phi}(A_j)$ and $\tilde{\Phi}(A_j^{-1} \ \beta \ A_j^{-1}) = \Phi(A_j^{-1})\tilde{\Phi}(\beta)\Phi(A_j^{-1})$,

for any $\beta \in \mathcal{B}$ and $j \in \mathbb{N}$.

For simplicity of notation, we sometimes write $\tilde{\Phi} = \Phi$ and call $\tilde{\Phi}$ the normal lifting of Φ to $\mathcal{B}$. Let us recall that $A_j \ \beta \ A_j$ denotes the sesquilinear form $(x, y) \in \mathcal{D} \times \mathcal{D} \to \beta(A_j x, A_j y)$.

Proof. If A is a linear operator in $\mathcal{B}$, with $A \geq \mathrm{Id}$ and $A\mathcal{D} = \mathcal{D}$ then $\bar{A}(\mathrm{Dom}\ \bar{A}) = H$, so that $\bar{A}$ is self-adjoint and $\bar{A}^{-1}(H) = \mathrm{Dom}(\bar{A})$. Taking A, B in $\mathcal{B}$ with $A(\mathcal{D}) = \mathcal{D}$, $B(\mathcal{D}) = \mathcal{D}$, we know from [12] p. 333 that the relation

$$\|Ax\| \geq \|Bx\| \geq \| x \|$$

for x in $\mathcal{D}$ implies that

$$\|\bar{A}^{-1}x\| \leq \|\bar{B}^{-1}x\| \leq \| x \|$$

for every $x \in \mathcal{D}$, hence

$$\Phi(\bar{A}^{-2}) \leq \Phi(\bar{B}^{-2}) \leq \Phi(1)$$

and, in the same manner (assuming that $\Phi(A^{-1})$, $\Phi(B^{-1})$ are injective)

$$\| \Phi(\bar{A}^{-1})^{-1} x\| \geq \| \Phi(\bar{B}^{-1})^{-1}x\| \geq \|x\|$$

for all $x \in \mathrm{Dom}\,(\bar{A}^{-1})^{-1}$, showing that the linear subset $\mathcal{D}'$ does not depend on the choice of the cofinal sequence A_i.

Fron structure of normal homomorphisms in von Neumann algebras, one has to treat the case where Φ is an ampliation, or an induction, the case of spatial isomorphism being obvious. Let Φ be an ampliation, and I be an index set such that the Hilbert space K coïncides with the direct sum $\oplus_{i\in I} H_i$ of copies H_i of copies H_i of H. We choose, with obvious notations $\tilde{\Phi}(A_i) = \oplus_{i\in I} A_i$, and the operators $\tilde{\Phi}(A_i)$ and $\tilde{\Phi}(A_i^{-1})$ are inverse w.r.t. each other on the dense linear subset $\mathcal{D}' = \cap_{i\geq 0} \mathrm{Dom}\tilde{\Phi}(A_i)$ of the Hilbert space K (paragraph 3, 3°/). For $\beta \in \mathcal{B}$, $\beta = B\, A_i \times A_i$, we define $\tilde{\Phi}(\beta) = \Phi(B)\tilde{\Phi}(A_i) \times \tilde{\Phi}(A_i)$. The sesquilinear form $\tilde{\Phi}(A_j)\tilde{\Phi}(\beta)\tilde{\Phi}(A_j)$ corresponds to the map

$$(x, y) \in \mathcal{D}'\times\mathcal{D}' \to (\Phi(B)\tilde{\Phi}(A_i)\tilde{\Phi}(A_j)x,\ \tilde{\Phi}(A_i)\tilde{\Phi}(A_j)y)$$

$$= (\Phi(B)\tilde{\Phi}(A_iA_j)x,\ \tilde{\Phi}(A_iA_j)y),$$

the calculation being the same for $\Phi(A_j^{-1})\tilde{\Phi}(\beta)\Phi(A_j^{-1})$, which is property 2°/.

Now, let Φ be an induction. Let E be some projector in M'; then $E\,\mathcal{D} \subset \mathcal{D}$, $K = E(H)$, hence $\mathcal{D}' = E(\mathcal{D})$ is dense in K. We choose here $\tilde{\Phi}(A_i) = A_iE$, hence $\mathcal{D}'$ can be identified to a closed subspace of the Fréchet space $\mathcal{D}$. For $\beta \in B(\mathcal{D}, \mathcal{D})$, we take $\tilde{\Phi}(\beta)$ equal to $\tilde{\Phi}(\beta)(x, y) = \beta(Ex, Ey)$ for $x, y \in E(\mathcal{D})$ and properties 2°/ are obvious.

It remains to show the normality of $\tilde{\Phi}$ and its uniqueness. The correspondence $B \in M \to B\, A_i \times A_i \in \mathcal{B}$ exchanges bounded increasing nets and their least upper bound and $\mathcal{B}_{A_i^2} = MA_i \times A_i$ from theorem 1.1; this gives the normality of $\tilde{\Phi}$, and hence its ultraweak continuity. Any normal linear positive map Ψ extending Φ being ultraweakly continuous must coïncide with $\tilde{\Phi}$ on $\mathcal{B}$ by theorem 1.2.

We point out that an isomorphism Φ from M onto itself does not necessarily induces a lifting $\tilde{\Phi}$ from $\mathcal{B}$ into itself.

Proposition 7.1. Let U be an unitary operator in M, and $\Phi(T) = U\,T\,U^{-1}$ be the corresponding isomorphism of M. Let $\tilde{\Phi}$ be the lifting of Φ to $\mathcal{B}$. Then

1°/ $\tilde{\Phi}(\mathcal{B}) \subset \mathcal{B}$ if an only if $U^{-1}\mathcal{D} \subset \mathcal{D}$

2°/ if U commute with all A_i^{-1}, one has $\tilde{\Phi}(M\,A_i \times A_i) = M\,A_i \times A_i$.

Of course, $U^{-1}\mathcal{D} = \mathcal{D}$ is equivalent to $\tilde{\Phi}(\mathcal{B}) = \mathcal{B}$, by 1°/ and the second assertion is always realized at the central sequence A_i.

Proof. It is immediate that $\tilde{\Phi}(A_i) = U\,A_i\,U^{-1}$, with $\mathcal{D}' = U(\mathcal{D})$. Let $\beta \in \mathcal{B}$ $B \in M$ and $i \in \mathbb{N}$ such that $\beta = B\,A_i \times A_i$. Then, for x, y in $\mathcal{D}'$,

$$\begin{aligned}\tilde{\Phi}(\beta)(x, y) &= (\Phi(B)\tilde{\Phi}(A_i)x, \tilde{\Phi}(A_i)y)\\ &= (U\,B\,U^{-1}\,U\,A_i\,U^{-1}x, U\,A_i\,U^{-1}y)\\ &= (B\,A_i\,U^{-1}\,x, A_i\,U^{-1}\,y),\end{aligned}$$

i.e., symbolically, $\tilde{\Phi}(\beta) = \beta(U^{-1}\,.\,, U^{-1}\,.\,) \in B(\mathcal{D}', \mathcal{D}')$. In particular, when U commute with all A_i^{-1}, we get $\tilde{\Phi}(\beta)(x, y) = (U\,B\,U^{-1}\,A_ix, A_iy)$; indeed, $U(\bar{A}_i^{-1}(H)) \subset \bar{A}_i^{-1}H$ leads to $U\,\mathcal{D} \subset \mathcal{D}$ and $U^* = U^{-1}$ also commute with A_i^{-1}, gives $U(\mathcal{D}) = \mathcal{D}$ and $U\,A_i = A_i\,U$ on $\mathcal{D}$, proving the second assertion.

Let us now assume that $U^{-1}\mathcal{D} \subset \mathcal{D}$. Then $A_i\,U^{-1}\mathcal{D} \subset \mathcal{D}$ and, by the closed graph theorem, $A_i\,U^{-1}$ is linear continuous fron the Fréchet space $\mathcal{D}$ into itself. Thus, one can find k in $\mathbb{N}$ and $M < +\infty$ such that $\|A_iU^{-1}x\| \leq M\|A_kx\|$ for all x in $\mathcal{D}$, therefore there exists C in M such that $A_iU^{-1} = C\,A_k$ on $\mathcal{D}$. Hence,

$$\begin{aligned}(\tilde{\Phi}(\beta)x, y) &= (B\,A_i\,U^{-1}x, A_i\,U^{-1}y)\\ &= (B\,C\,A_kx, C\,A_ky) = (C^*\,B\,C\,A_kx, A_ky)\end{aligned}$$

and, from $C^*BC \in M$, we show that $\tilde{\Phi}(\beta) \in \mathcal{B}$, i.e. $\tilde{\Phi}(\mathcal{B}) \subset \mathcal{B}$.

Conversely, let us start from $\tilde{\Phi}(\mathcal{B}) \subset \mathcal{B}$. With β of the form $\beta = A_i \times A_i$, we get $\tilde{\Phi}(\beta) = A_i U^{-1} \times A_i U^{-1} \in \mathcal{B}$: since this element is dominated by some A_j^2, there exists M such that

$$\|A_i\ U^{-1}x\| \le M\|A_j x\|$$

for $x \in \mathcal{D}$, hence

$$\|\overline{A_i\ U^{-1}}x\| \le M\|\bar{A}_j x\|$$

for $x \in \mathrm{Dom}(\bar{A}_j)$. But $\bar{A}_j(\mathrm{Dom}\ \bar{A}_j) = H$ and $\overline{A_i\ U^{-1}} = \bar{A}_i\ \bar{U}^{-1}$ imply that

$$\|\bar{A}_i\ \bar{U}^{-1}\ \bar{A}_j^{-1}x\| \le \|x\|$$

for x in H, hence $U^{-1}\mathcal{D} \subset \mathrm{Dom}(\bar{A}_i)$. Since this holds for every i, we get $U^{-1}\mathcal{D} \subset \mathcal{D}$.

It is instructive to consider

Proposition 7.2. Let A be an ultraweakly closed $*$-algebra, with a cofinal central sequence, f be a positive normal linear form on A, and f_1 be its restriction to A_{id}. We denote by $(\pi_f, \zeta_f, \mathcal{D}_f, H_f)$ (resp. (π_1, ζ_1, H_1)) the G.N.S. representation of the $*$-algebra A (resp. A_{id}). Then:

1°/ there exists, between Hilbert spaces H_1 and H_f, a unitary operator $U : H_f \to H_1$, which exchanges the restriction of π_f to A_{id} with π_{f_1}, and which satisfies $U(\zeta_f) = \zeta_1$

2°/ the representation π_f is the normal lifting of π_{f_1}, and $\pi_f(A)$ is an ultraweakly closed $*$-algebra.

Proof. We denote by $\pi_{f|A_{id}}$ the representation of A_{id} obtained by restricting π_f to A_{id}, and by j the canonical injection from A into H_f; then $\zeta_f = j(1)$ and $\mathcal{D}_f$ is, by definition, the Fréchet space completion of the linear subset $j(A)$. We first show that ζ_f is a cyclic vector for $\pi_{f|A_{id}}$; indeed, j is continuous when A is endowed with ρ-topology because

$$\|\pi_f(T)\zeta_f\|^2 = (\pi_f(T^*T)\zeta_f, \zeta_f) = f(T^*T)$$

tends to zero when T tends to zero for λ-topology (and $\lambda = \rho$ by corollary 6.1) and A_{id} is λ-dense in A by theorem 2.1. Now, one has, for $T \in A_{id}$,

$$(\pi_f(T)\zeta_f, \zeta_f) = f(T) = f_1(T) = (\pi_1(T)\zeta_1, \zeta_1).$$

Since ζ_1 is cyclic for π_1, the representations π_{f_1} and $\pi_f|A_{id}$ are unitarily equivalent, which is 1°/.

Clearly, f_1 is positive normal on A_{id}, therefore π_{f_1}, hence $\pi_f|A_{id}$ are positive normal homomorphisms. Using the fact that the map $T \in A_{id} \to A_i\ T\ A_i = T\ A_i \times A_i \in A_{A_i^2}$ is bijective and normal and that $\pi_f(A_i\ T\ A_i) = \pi_f(A_i)\pi_f(T)\pi_f(A_i)$, it follows that π_f is normal, hence ultraweakly continuous on A. Now, define π' to be the normal lifting, of π, and $\mathcal{D}'$ be the corresponding domain of $\pi'(A)$. We need to show that $\mathcal{D}_f = \mathcal{D}'$. We know that $\pi_f(A_i^{-1})\mathcal{D}_f = \mathcal{D}_f$, $\pi_f(A_i)\mathcal{D}_f = \mathcal{D}_f$ $\pi'(A_i^{-1})\mathcal{D}' = \mathcal{D}'$, $\pi'(A_i)\mathcal{D}' = \mathcal{D}'$ with, moreover, $\pi'(A_i^{-1}) = \pi_f(A_i^{-1})$. Thus, $\text{Dom}(\overline{\pi_f(A_i)}) = \pi_f(A_i^{-1})(H_f)$, $\text{Dom}\ \overline{\pi'(A_i)} = \pi'(A_i^{-1})(H_f)$, hence $\mathcal{D}_f = \mathcal{D}'$. Finally, π and π' are ultraweakly continuous, and hence they coïncide on A by theorem 1.2. In particular, intervals $[-\pi_f(A_i), \pi_f(A_i)]$ are ultraweakly compact, showing that $\pi_f(A)$ is an ultraweakly closed *-algebra in $B(\mathcal{D}_f, \mathcal{D}_f)$.

<u>Definition 7.1</u>. Let A be a subspace of $B(\mathcal{D}, \mathcal{D})$, with cofinal sequence A_i, satisfying condition II, and H be the Hilbert space completion of $\mathcal{D}$. Let $\mathcal{B}$ be a similar space, with natural domain $\mathcal{D}'$, acting in some Hilbert space K (in general, $K \neq H$). A linear map π from A into $\mathcal{B}$ is called an <u>homomorphism, or a representation</u>, iff the following properties are satisfied:

1°/ π is a linear positive map, $\pi(A_i)\mathcal{D}' = \mathcal{D}'$, $\pi(A_i^{-1})\mathcal{D}' = \mathcal{D}'$, $\pi(A_i)^{-1} = \pi(A_i^{-1})$, $\pi(A_i\ A_k^{-1}) = \pi(A_i)\pi(A_k^{-1})$ for $i, k \in \mathbb{N}$, and the sequence $\pi(A_i)$ is cofinal in $\mathcal{B}^+$.

2°/ The restriction of π to A_{id} is a homomorphism (as defined in [4]) from the involutive algebra A_{id} into the involutive algebra $\mathcal{B}_{id}$ (with $\pi(1) = 1$).

3°/ For $\beta \in \mathcal{A}$ and $i \in \mathbb{N}$, one has $\pi(A_i^{-1})\pi(\beta)\pi(A_i^{-1}) =$ $= \pi(A_i^{-1}\ \beta\ A_i^{-1})$ and $\pi(A_i)\pi(\beta)\pi(A_i) = \pi(A_i\ \beta\ A_i)$.

Since positive linear maps decrease ρ-norms, a representation π from $\mathcal{A}$ into $\mathcal{B}$ is continuous for respective ρ-topologies.

Lemma 7.1. We put

3'/ for every β in $\mathcal{A}$, there exists an integer i such that $A_i^{-1}\ \beta\ A_i^{-1}$ is bounded with $\pi(\beta) = \pi(A_i)\pi(A_i^{-1}\ \beta\ A_i^{-1})\pi(A_i)$

3"/ for every β in $\mathcal{A}$ such that $A_i^{-1}\ \beta\ A_i^{-1}$ is bounded, one has $\pi(\beta) = \pi(A_i)\pi(A_i^{-1}\ \beta\ A_i^{-1})\pi(A_i)$.

When 1°/ and 2°/ are satisfied, then the formulations 3, 3', 3" are equivalent.

Proof. It is clearly seen that 3/ $\Rightarrow$ 3"/ $\Rightarrow$ 3'/. Now, assume that 3' is true. Let $\beta \in \mathcal{A}$, and i, j integers such that $A_i^{-1}\ \beta\ A_i^{-1}$ and $A_j^{-1}\ \beta\ A_j^{-1}$ are bounded. Since $\mathcal{A}$ satisfies condition II, there exists B_i, B_j in $\mathcal{A}_{id}$ such that $\beta = B_i\ A_i \times A_i = B_j\ A_j \times A_j$, with $\pi(\beta) = \pi(B_i)\pi(A_i) \times \pi(A_i)$ by 3'. The operator B_j is bounded and coïncides with $(A_i\ A_j^{-1})^* B_j (A_i\ A_j^{-1})$ on $\mathcal{D}$. We choose the integer k $(k \geq i,\ k \geq j)$ such that $A_i\ A_k^{-1}$ are bounded; then $(A_i\ A_k^{-1})^*$ and $(A_j\ A_k^{-1})^*$ are bounded, with adjoints equal to $A_k^{-1}\ A_i$ and $A_k^{-1}\ A_j$ on $\mathcal{D}$. Taking u, v in $\mathcal{D}'$, one has, by 1°/,

$$(\pi(B_j)\pi(A_j)u,\ \pi(A_j)v) = (\pi(B_j)\pi((A_j\ A_r^{-1})\ A_r)u,\ \pi(A_j\ A_r^{-1})\ A_r)v)$$

equal (by 2°/) to the following three expressions:

$$(\pi(B_j)\pi(A_j\ A_r^{-1})\pi(A_r)u,\ \pi(A_r^{-1}\ A_j)^*\pi(A_r)v)$$
$$\text{(due to } \pi(A_j^{-1})\pi(A_j\ A_r^{-1}) = \pi(A_r^{-1})\text{)},$$

$$(\pi(A_r^{-1}\ A_j)\pi(B_j)\pi(A_j\ A_r^{-1})\pi(A_r)u,\ \pi(A_r)v),$$

and

$$(\pi(\overline{A_r^{-1}\ A_j\ B_j}\ A_j\ A_r^{-1})\pi(A_r)u, \pi(A_r)v).$$

In the same way, replaining j by i,

$$(\pi(B_i)\pi(A_i)\ ,\ \pi(A_i)v) = (\pi(\overline{A_r^{-1}\ A_i}\ B_iA_iA_r^{-1})\pi(A_r)u, \pi(A_r)v),$$

or, for u, v $\in \mathcal{D}$, noting that $A_k\mathcal{D} = \mathcal{D}$, we get

$$(B_j\ A_j\ A_k^{-1}u,\ A_j\ A_k^{-1}v) = (B_i\ A_i\ A_j^{-1}\ A_j\ A_r^{-1}u,\ A_i\ A_j^{-1}\ A_j\ A_r^{-1}v)$$

$$= (B_i\ A_i\ A_r^{-1}u,\ A_i\ A_r^{-1}v).$$

Therefore, for u $\in \mathcal{D}$,

$$\overline{A_k^{-1}\ A_j}\ B_j\ A_j\ A_k^{-1}u = \overline{A_k^{-1}\ A_i}\ B_i\ A_i\ A_k^{-1}\ u.$$

Since these operators are bounded, it follows that $\pi(B_i)\pi(A_i) \times \pi(A_i) = \pi(B_j)\pi(A_j) \times \pi(A_j)$, which is 3".
To prove 3/, let $\beta \in A$, and $k \in \mathbb{N}$, $B \in A_{id}$ such that $\beta = B\ A_k \times A_k$; as seen before, $\pi(\beta) = \pi(B)\pi(A_k) \times \pi(A_k)$.
Let i be any integer, and j chosen in such a way that $A_k\ A_i$ is smaller than some homothetic of A_j, then one can find $C \in A_{id}$, such that $A_k\ A_i = C\ A_j$.
Clearly, $A_i\ \beta\ A_i = B\ A_k\ A_i \times A_k\ A_i = C^*\ B\ C\ A_j \times A_j$, hence, by 3",

$$\pi(A_i\ \beta\ A_i) = \pi(C^*)\pi(B)\pi(C)\pi(A_j) \times \pi(A_j).$$

Replacing C by $A_k\ A_i\ A_j^{-1}$, and noting that $\pi(A_i) = \pi(A_i\ A_j^{-1})\pi(A_j)$, we get $\pi(A_k^{-1}\ C) = \pi(A_i\ A_j^{-1})$, hence $\pi(C) = \pi(A_k)\ \pi(A_i)\ \pi\ (A_j^{-1})$, leading to $\pi(A_i\ \beta\ A_i) = \pi(A_i)\pi(\beta)\pi(A_i)$. The proof is similar, with A_i^{-1} in place of A_i, thus proving the lemma.

<u>Proposition 7.3</u>. Take π, A, B, ... of definition 7.1, A being a $*$-algebra satisfying condition I. Then, $\pi(A)$ is a $*$-algebra with natural domain $\mathcal{D}'$, satisfying condition I, and π is an involutive homomorphism for the respective algebraic structures of A and $\pi(A)$.

<u>Proof</u>. Condition I imply condition II, hence $T \to T\ A_i \times A_i = A_i\ T\ A_i$ exchanges A_{id} and $A_{A_i^2}$. We first show that elements T of $\pi(A_{id})$ send $\mathcal{D}'$ into itself. From $T\ \mathcal{D} \subset \mathcal{D}$ follows the

continuity of T viewed as a map from the Fréchet space $\mathcal{D}$ into itself; hence, for every i, there exists j and $M < +\infty$, such that

$$\|A_i \; Tx\| \leq M\|A_j x\|$$

for all $x \in \mathcal{D}$, showing that $A_i \; T \; A_j^{-1}$ is bounded. Clearly $T = A_i^{-1} \; A_i \; T \; A_j^{-1} \; A_j$ on $\mathcal{D}$, hence $\pi(T) = \pi(A_i^{-1})\pi(A_i \; T \; A_j^{-1})\pi(A_j)$ on $\mathcal{D}'$, by 2°/. Recalling that $\pi(A_i^{-1})K = \text{Dom}\;\overline{\pi(A_i)}$, we get, for x in $\mathcal{D}'$,

$$\pi(T)\pi(A_j^{-1})x = \pi(A_i^{-1})\pi(A_i \; T \; A_j^{-1})x$$

hence $\pi(T)\pi(A_j^{-1})x \in \text{Dom}(\overline{A_i})$, and from $\pi(A_j^{-1})\mathcal{D} = \mathcal{D}$, we get $\pi(T)\mathcal{D}' \subset \text{Dom}\;(\overline{A_i})$: we deduce $\pi(T)\mathcal{D}' \subset \mathcal{D}'$ since this holds for every i. It is now an easy computation that $\pi(T)^* = \pi(T^*)$, $\pi(ST) = \pi(S)\pi(T)$, for $S, T \in \mathcal{A}$.

Another observation is

<u>Lemma 7.2</u>. Let $\mathcal{A}$ be a $\star$-algebra satisfying condition I, f a positive linear form on $\mathcal{A}$, and $(\pi_f, H_f, \mathcal{D}_f, \zeta_f)$ be the G.N.S. representation of $\mathcal{A}$, associated to f, with cyclic vector $\zeta_f (\zeta_f \in \mathcal{D}_f)$. Then, in the Hilbert space H_f, ζ_f is a cyclic vector for $\pi_f(\mathcal{A}_{id})$.

<u>Proof</u>. By the definition of G.N.S. construction, the linear set $\pi_f(\mathcal{A})\zeta_f$ is dense in H_f. We denote by K the closure in H_f of the linear set $\pi_f(\mathcal{A}_{id})\zeta_f$. For $B \in \mathcal{A}_{id}$, one clearly has $B\mathcal{A}_{id} \subset \mathcal{A}_{id}$, hence $\pi_f(B)K \subset K$, so that the restriction π of π_f to K defines a representation of the involutive algebra $\mathcal{A}_{id}$. Let $\pi^{\perp}$ be the representation of $\mathcal{A}_{id}$ obtained by restricting π_f to $H_f \ominus K$ (i.e., $\pi_f = \pi \oplus \pi^{\perp}$ on $\mathcal{A}_{id}$). Take $v \in \pi_f(\mathcal{A})\zeta_f$ and $T \in \mathcal{A}$ such that $v = \pi_f(T)\zeta_f$. There exists an integer i such that $T^* A_i^{-1} \in \mathcal{A}_{id}$; the adjoint of this operator is obviously bounded, and coincides on $\mathcal{D}$ with $A_i^{-1}\; T$, hence $A_i^{-1} \; T \in \mathcal{A}_{id}$.
Now,

$$\pi_f(A_i^{-1} \; T)\zeta_f = \pi_f(A_i^{-1})\pi_f(T)\zeta_f = \pi_f(A_i^{-1})v$$

shows that $\pi_f(A_i^{-1})v \in K \cap \mathcal{D}_f$. Now, $\pi_f(A_i)\mathcal{D}_f = \mathcal{D}_f$, $\pi_f(A_i^{-1})\mathcal{D}_f = \mathcal{D}_f$ and, since $\pi_f(A_i^{-1})$ commutes with the projector p_K associated to K, we obtain $p_K(\mathcal{D}_f) \subset \mathcal{D}_f$ and $\pi_f(A_i)p_K = p_K\pi_f(A_i)$ on $\mathcal{D}_f$. Hence $v = \pi_f(A_i)\pi_f(A_i^{-1})v \in K$, so that $\pi_f(\mathcal{A})\zeta_f \subset K$, which is the lemma.

<u>Lemma 7.3</u>. Let $\mathcal{A}$ be a space satisfying condition II with cofinal sequence A_i, $\mathcal{B}_1$ (resp. $\mathcal{B}_2$) be similar objects with natural domain $\mathcal{D}_1$(respectively $\mathcal{D}_2$), acting in Hilbert spaces K_1 (respectively K_2) and π_1 (respectively π_2) be a homomorphism from $\mathcal{A}$ into $\mathcal{B}_1$ (respectively $\mathcal{B}_2$). We assume that there exists $x_1 \in K_1$ (respectively $x_2 \in K_2$) cyclic for $\pi_1(\mathcal{A}_{id})$ (respectively $\pi_2(\mathcal{A}_{id})$), such that one has $(\pi_1(T)x_1, x_2) = (\pi_2(T)x_2, x_2)$ for $T \in \mathcal{A}_{id}$. Then, there exists a unitary operator U from K_1 onto K_2, which exchanges $\mathcal{D}_1$ and $\pi_2(\mathcal{A})$), with $(\pi_1(\beta)u, v) = (\pi_2(\beta)Uu, Uv)$ for $u, v \in \mathcal{D}_1$ and $\beta \in \mathcal{A}$.

The operator U, as constructed, satisfies $Ux_1 = x_2$, hence we see that $x_1 \in \mathcal{D}_1$ implies that $x_2 \in \mathcal{D}_2$.

<u>Proof</u>. It is immediate that the relation $\|\pi_1(T)x_1\| = \|\pi_2(T)x_2\|$, for $T \subset \mathcal{A}_{id}$, gives an unitary operator U from K_1 onto K_2 characterized by $U(\pi_1(T)x_1) = \pi_2(T)x_2$, for $T \in \mathcal{A}_{id}$. One has $\pi_2(T)U = U\pi_1(T)$ for bounded T, hence with $T = A_i^{-1}$ we get $U(\pi_1(A_i^{-1})K_1) = \pi_2(A_i^{-1})K_2$, i.e., $U\mathcal{D}_1 = \mathcal{D}_2$. Since $\pi_1(A_i)$ respectively $\pi_2(A_i)$) is exactly the inverse on $\mathcal{D}_1$ (respectively $\mathcal{D}_2$) of $\pi_1(A_i^{-1})$ (respectively $\pi_2(A_i^{-1})$), we deduce that $\pi_2(A_i)U = U\pi_1(A_i)$ on $\mathcal{D}_1$. Taking $\beta = B\, A_i \times A_i$ in $\mathcal{A}$, we get

$$\begin{aligned}\pi_1(\beta) &= \pi_1(B)\pi_1(A_i) \times \pi_1(A_i)\\ &= \pi_1(B)U^{-1}\,U\,\pi_1(A_i) \times U^{-1}\,U\,\pi_1(A_i)\\ &= \pi_2(B)\pi_2(A_i)U \times \pi_2(A_i)U = U^*\pi_2(\beta)U,\end{aligned}$$

hence the lemma.

Proposition and definition 7.4. Let $\mathcal{B}$ be an ultraweakly closed space satisfying condition II, and f be a positive normal linear form on $\mathcal{B}$, with restriction f_1 to the von Neumann algebra $\mathcal{B}_{id}$. Let (π_1, H_1, ζ_1) be the usual G.N.S. representation of $\mathcal{B}_{id}$, associated to f_1. The normal lifting π_f (with domain $\mathcal{D}_f$) of π_1 to $\mathcal{B}$ is called the G.N.S. representation of $\mathcal{B}$ associated to f, and has following properties:

1°/ π_f is a representation of $\mathcal{B}$ (in sense of definition 7.1)

2°/ ζ_1 belongs to $\mathcal{D}_f$, and is a cyclic vector for $\pi_f(\mathcal{B}_{id})$

3°/ for $\beta \in \mathcal{B}$, one has $f(\beta) = (\pi_f(\beta)\zeta_1, \zeta_1)$

By lemma 7.3, properties 1°/, 2°/, 3°/ characterize G.N.S. construction (up to a unitary operator exchanging domain). It is now natural to write ζ_f instead of ζ_1.

Proof. Let A_i be a cofinal sequence in $\mathcal{B}$ with $\mathcal{D} = \cap_{i\geq 0} A_i^{-1}(H)$, where H is the Hilbert space completion of $\mathcal{D}$. By definition, $\mathcal{D}_f = \cap_{i\geq 0}\, \pi_1(A_i^{-1})(H_f)$, and π_f coïncides with π_1 on $\mathcal{B}_{id}$. Let us choose an integer i, and write $\Delta = A_i$ for simplicity. The second assertion will be shown as soon as the relation $\zeta_1 \in \mathrm{Dom}(\overline{\pi_f(\Delta)}) = \pi_1(\Delta^{-1})H_1$ is established. Let $\mathcal{C}$ be the $*$-algebra generated by Δ, Δ^{-1}, with natural domain $\mathcal{D}_\Delta$: clearly, $\mathcal{D}_\Delta \supset \mathcal{D}$ and $\mathcal{D}_\Delta = \cap_{k\geq 0}\Delta^{-k}(H)$. The von Neumann algebra generated by $\mathcal{C}_{id}$ is contained in $\mathcal{B}_{id}$; therefore, the σ-weak closure $\mathcal{C}^\sigma$ of $\mathcal{C}$, relative to $\mathcal{D}_\Delta \hat{\otimes} \mathcal{D}_\Delta$, is an abelian $*$-algebra contained in $\mathcal{B}$. Let us define, in the same way, $(\mathcal{D}_f)_\Delta = \cap_{k\geq 0}\pi_1(\Delta^{-k})(H_1)$, and note that π_f, restricted to $\mathcal{C}^\sigma$, is a representation of $\mathcal{C}^\sigma$ on the domain $(\mathcal{D}_f)_\Delta$. Since $\mathcal{D}_\Delta$ is essentially dense, there exists an increasing sequence p_n $(n\geq 2)$ of projectors in $\mathcal{C}^\sigma_{id}$ such that $p_n(H) \subset \mathcal{D}_\Delta$ and $(\cup_n\, p_n(H))^\perp = 0$. By the closed graph theorem, the operators $\Delta^k p_n$ are bounded, for every $k \in \mathbb{N}$, in $\mathcal{C}^\sigma_{id}$ so that $\pi_1(\Delta^k\, p_n)$ are also bounded in H_1. From $\pi_1(p_n) = \pi_1(\Delta^{-k})\pi_1(\Delta^k\, p_n)$ follows $\pi_1(p_n)H_1 \subset \mathrm{Dom}\overline{\pi_f(\Delta)^k}$, hence $\pi_1(p_n)H_1 \subset (\mathcal{D}_f)_\Delta$ for every n and $\pi_f(\Delta^k)\pi_1(p_n) = \pi_1(\Delta^k p_n)$. We put $\zeta_n = \pi_1(p_n)\zeta_1 \in (\mathcal{D}_f)_\Delta$, and obtain

$$\|\zeta_n - \zeta_1\|^2 = \|\pi_1(p_n)\zeta_1 - \pi_1(\mathrm{Id})\zeta_1\|^2$$
$$= \|\pi_1(p_n - \mathrm{Id})\zeta_1\|^2$$
$$= f((p_n - \mathrm{Id})^2),$$

which tends to zero, since projectors p_n increase to Id and f is normal on $\mathcal{B}$. Now, the quantity

$$\|\pi_f(\Delta)\pi_1(p_n)\zeta_1 - \pi_f(\Delta)\pi_1(p_m)\zeta_1\|^2$$

makes sense, and coïncides with

$$\|\pi_1(\Delta p_n - \Delta p_m)\zeta_1\|^2 = f(\Delta^2(p_n - p_m)).$$

However, $\Delta^2 p_n = p_n\Delta \times \Delta$ increases to Δ^2, and normality of f implies that $\pi_f(\Delta)\zeta_n$ is a Cauchy sequence in H; hence $\zeta_1 \in \mathrm{Dom}(\overline{\pi_f(\Delta)})$. The first assertion is shown in theorem 7.1, and π_f is ultraweakly continuous from $\mathcal{B}$ in $B(\mathcal{D}_f, \mathcal{D}_f)$. Finally, relation 3°/ is immediate for $\beta \in \mathcal{B}_{id}$: each side being an ultraweakly continuous function of β (due to $\zeta \in \mathcal{D}_f$), the equality remains true on the whole $\mathcal{B}$ by theorem 1.2.

Remark 7.1. Care should be taken with G.N.S. construction. Indeed, let $g \geq 0$ be a normal linear functional on the von Neumann algebra $\mathcal{B}_{id}$, (π_g, H_g, ζ_g) be the G.N.S. representation of $\mathcal{B}_{id}$, with cyclic vector ζ_g, and π be its normal lifting to $\mathcal{B}$. Then π is not necessarily the G.N.S. representation associated with some positive normal linear functional on $\mathcal{B}$ (see lemma 7.7 and theorem 7.2).

Proposition 7.5. Let $\mathcal{B}$ be an ultraweakly closed space satisfying condition II with cofinal sequence A_i, and π a normal homomorphism, with domain $\mathcal{D}_\pi$ from $\mathcal{B}$ into $B(\mathcal{D}_\pi, \mathcal{D}_\pi)$. Then, $\pi(\mathcal{B})$ is an ultraweakly closed space of $B(\mathcal{D}_\pi, \mathcal{D}_\pi)$, satisfying condition II, and $\pi(\mathcal{B}_{A_i}) = \pi(\mathcal{B})_{A_i}$ for every i (in particular $\pi(\mathcal{B}_{id}) = \pi(\mathcal{B})_{id}$).

Proof. Since $\mathcal{B}$ is stable under maps $\beta \to A_i\ \beta\ A_i$ and $\beta \mapsto A_i^{-1}\ \beta\ A_i^{-1}$, $\pi(\mathcal{B})$ has the same properties relative to $\pi(A_i)$ and $\pi(A_i^{-1})$, due to 3' of definition 7.1. The restriction of π to $\mathcal{B}_{id}$ is a normal homomorphism of $\mathcal{B}_{id}$, hence $\pi(\mathcal{B}_{id})$ is a von Neumann algebra.

From $\mathcal{B} = \cup_i \mathcal{B}_{id} A_i \times A_i$ it follows that $\pi(\mathcal{B}) = \cup_i \pi(\mathcal{B}_{id})\pi(A_i) \times \pi(A_i)$. Clearly, $\pi(A_i^{-1}) \in \pi(\mathcal{B}_{id})$, therefore theorem 1.1 shows that $\pi(\mathcal{B})_{id} = \pi(\mathcal{B}_{id})$ and $\pi(\mathcal{B})$ satisfy condition II. For every positive normal linear form φ on $\pi(\mathcal{B})$, $\varphi \circ \pi$ is a positive normal linear form on $\mathcal{B}$ hence π is ultraweakly continuous, which obviously implies that $\pi(\mathcal{B})$ is ultraweakly closed. The relation $\pi(\mathcal{B}_{A_i}) = \pi(\mathcal{B})_{A_i}$ follows from lemma 1.1, since $A_i^{-\frac{1}{2}}(\mathcal{D}) = \mathcal{D}$ and $A_i^{-\frac{1}{2}} \in \mathcal{B}$.

Proposition 7.6. Let A be a $\star$-algebra satisfying condition I, with natural domain $\mathcal{D}$, $\mathcal{B}$ be its σ-weak closure relative to $\mathcal{D} \hat{\otimes} \mathcal{D}$. Let f be given an ultraweakly continuous positive linear form on A, and let g be its unique ultraweakly continuous extension to $\mathcal{B}$. Then, the G.N.S. representation $(\pi_f, H_f, \mathcal{D}_f, \zeta_f)$ of the $\star$-algebra A, associated to f, is ultraweakly continuous, and the unique ultraweakly continuous extension of π_f to $\mathcal{B}$ coïncides with the G.N.S. representation of $\mathcal{B}$, which is associated to g.

Proof. Let $(\pi_g, H_g, \mathcal{D}_g, \zeta_g)$ be the G.N.S. representation of $\mathcal{B}$, associated to g. As π_g is normal on $\mathcal{B}$, its restriction to $\mathcal{B}_{id}$ is Hilbert-ultraweakly continuous, hence $\pi_g(A_{id})$ is Hilbert-ultraweakly dense in the von Neumann algebra $\pi(\mathcal{B}_{id})$, due to $\mathcal{B}_{id} = (A_{id})''$, see [1] lemma 8.5. In particular, $\pi(A_{id})$ is a strongly dense $\star$-subalgebra of $\pi(\mathcal{B}_{id})$, so that the closed linear spans of $\pi_g(A_{id})\zeta_g$ and $\pi_g(\mathcal{B}_{id})\zeta_g$ in the Hilbert space H_g are identical, hence ζ_g is cyclic for $\pi_g(A_{id})$. Now, by lemma 7.2, ζ_f is cyclic for $\pi_f(A_{id})$, and by lemma 7.3 there exists a unitary operator U from H_g onto H_f such that $U\zeta_g = \zeta_f$, and $\pi_g(\beta) = U^*\pi_f(\beta)U$ for β in A, due to

$$f(T) = (\pi_g(T)\zeta_g, \zeta_g) = (\pi_f(T)\zeta_f, \zeta_f)$$

for T in A_{id}. By definition, $\mathcal{D}_f = \cap_{i\geq 0}\pi_f(A_i^{-1})H_f$, $\mathcal{D}_g = \cap_{i\geq 0}\pi_g(A_i^{-1})H_g$, $U\mathcal{D}_g = \mathcal{D}_f$, so that π_f can be identified to π_g restricted to A (because U exchanges ultraweak topologies). Finally, π_g is ultraweakly continuous on $\mathcal{B}$, and hence on A_{id}, so that all assertions become straightforward.

Before continuing, one has to introduce subrepresentations, and direct sums of representations as one divines. Let A be a space satisfying condition II, and π be a representation of A, with domain $\mathcal{D}$; A_{id} being an involutive algebra of bounded operators and the restriction π_{id} of π to A_{id} being a representation (of A_{id}) in the Hilbert space H completion of $\mathcal{D}$ such that $\pi_{id}(1) = 1$. Since the operators $\pi(A_i)$ are essentially self-adjoint on $\mathcal{D}$, we know that $\mathcal{D} = \cap_{i\geq 0} \pi_{id}(A_i^{-1})(H)$ (the sequence A_i being obviously cofinal in A^+), and for $\beta = B\, A_i \times A_i$ one has $\pi(\beta) = \pi(B)\pi(A_i) \times \pi(A_i)$.

Let H_1 be a closed subspace of H, stable under $\pi(A_{id})$, and p_1 be the corresponding projector. Then, p_1 commutes with $\pi(A_{id})$, implying that $p_1(\mathcal{D}) \subset \mathcal{D}$ $(1 - p_1)\mathcal{D} \subset \mathcal{D}$, and hence that $\mathcal{D} = p_1\mathcal{D} \oplus (1 - p_1)\mathcal{D}$. By proposition 4.1, $\mathcal{D}_1 = p_1(\mathcal{D})$ is a closed subspace of the Fréchet space $\mathcal{D}$.

We introduce π_1 to be the representation of A_{id}, obtained by restricting elements of $\pi(A_{id})$ to H_1, and denote by $\pi_1(A_i)$ the operator $p_1\pi(A_i)$, with domain Dom $\pi_1(A_i) = \mathcal{D}_1$, acting in H_1.

Lemma 7.4. One has $\pi_1(A_i)\mathcal{D}_1 = \mathcal{D}_1$, $\pi_1(A_i^{-1})\mathcal{D}_1 = \mathcal{D}_1$, and $\pi_1(A_i^{-1})$, $\pi_1(A_i)$ are inverse w.r.t. each other on $\mathcal{D}_1$. The domain $\mathcal{D}_1$ is complete under topology defined by the sequence of semi-norms $x \mapsto \|\pi_1(A_i)x\|$.

It amounts to the same to say that $\mathcal{D}_1$ is the natural domain of the $\star$-algebra generated by all $\pi_1(A_i)$, $\pi_1(A_i^{-1})$.

Proof. The relations $\pi(A_i)p_1(x) = p_1\pi(A_i)x$, for x in $\mathcal{D}$, and $\pi(A_i^{-1})p_1 = p_1\pi(A_i^{-1})$, $\pi(A_i)\mathcal{D} = \mathcal{D}$ lead to $\pi(A_i)\mathcal{D}_1 \subset \mathcal{D}_1$ and $\pi(A_i)^{-1}\, \mathcal{D}_1 \subset \mathcal{D}_1$, hence $\pi_1(A_i)\mathcal{D}_1 = \mathcal{D}_1$, and $\pi_1(A_i)^{-1}$ is the inverse of $\pi_1(A_i^{-1})$ on $\mathcal{D}_1$. It follows that $\pi_1(A_i)$ is essentially self-adjoint on $\mathcal{D}_1$, hence $\mathcal{D}_1 \subset \cap_{i\geq 0}\, \pi_1(A_i^{-1})(H_1)$. Now, $\pi_1(A_i)\mathcal{D}_1$ is dense in H_1, so that $\mathcal{D}_1$ is a core of $\mathrm{Dom}(\overline{\pi_1(A_i)})$ for every i, which means that $\mathcal{D}_1$ is dense in the Fréchet space

$\cap_{i\geq 0}$ Dom $\overline{\pi_1(A_i)}$. However, $\mathcal{D}_1$ is complete for semi-norms $x \in \mathcal{D}_1 \to \|\pi(A_i)x\| = \|\pi(A_i)p_1x\|$, hence $\mathcal{D}_1 = \cap_{i\geq 0}$ Dom $\overline{\pi_1(A_i)}$. Finally, Dom $\overline{\pi_1(A_i)} = \pi_1(A_i^{-1})(H_1)$ since $\pi(A_i) \geq 1$ is essentially self-adjoint, hence the lemma.

From $\pi_1(A_i)\mathcal{D}_1 \subset H_1$, $\pi_2(A_i)\mathcal{D}_2 \subset H_2$, follows

Lemma 7.5. Let $p_2 = 1 - p_1$, $H_2 = p_2(H)$, $\mathcal{D}_2 = p_2(\mathcal{D})$ and $\beta \in A$. Then

$$(\pi(\beta)x, y) = 0 \qquad \text{for } x \in \mathcal{D}_1, \ y \in \mathcal{D}_2.$$

Now, for $\beta = B\, A_i \times A_i$, formulas $\pi_1(\beta) = \pi_1(B)\pi_1(A_i) \times \pi_1(A_i)$ and $\pi_2(\beta) = \pi_2(B)\pi_2(A_i) \times \pi_2(A_i)$ unambiguously define two representations of A, with their respective domains $\mathcal{D}_1$ and $\mathcal{D}_2$, which completely determine π. We shall write $\pi = \pi_1 \oplus \pi_2$.

More generally, let p_α, $\alpha \in I$ (I index set) be a family of projectors in $\pi(A_{id})'$, such that $\oplus_{\alpha\in I}\, p_\alpha = \mathrm{Id}$ in H. We put $H_\alpha = p_\alpha(H)$, $\mathcal{D}_\alpha = p_\alpha(\mathcal{D})$, and let $\mathcal{D}_0$ be the algebraic direct sum of all $\mathcal{D}_\alpha$. The operator $\pi_0(A_i)$, which coïncides on each $\mathcal{D}_\alpha$ with the operator $\pi_\alpha(A_i)$, clearly satisfies $\pi_0(A_i)\mathcal{D}_0 = \mathcal{D}_0$, $\pi_0(A_i^{-1})\mathcal{D}_0 = \mathcal{D}_0$, and hence is essentially self-adjoint on $\mathcal{D}_0$ and its closure coïncides with $\overline{\pi(A_i)}$. The domain $\mathcal{D}_\pi$ of π is, therefore, the natural domain of the $*$-algebra generated by all $\pi_0(A_i)$, $\pi_0(A_i^{-1})$. Now, $\mathcal{D}_\pi$ is known to be equal to the set of all σ-convergent families $x = (x_\alpha)_{\alpha\in I}$ with $x_\alpha \in \mathcal{D}_\alpha$ for all α. Finally, for $\beta = B\, A_i \times A_i$ in A, $x = (x_\alpha)_{\alpha\in I}$, $y = (y\ )_{\alpha\in I}$ in $\mathcal{D}$, one has

$$(\pi(\beta)x, y) = (\pi(B)\pi(A_i)x, \pi(A_i)y)$$

$$= \sum_{\alpha\in I} (\pi_\alpha(B)\pi_\alpha(A_i)x, \pi_\alpha(A_i)y),$$

the serie considered being absolutely convergent, so that one can write $\pi = \oplus_{\alpha\in I}\, \pi_\alpha$, where π_α is the representation of A in H_α with domain $\mathcal{D}_\alpha$, associated with the formula $\pi_\alpha(\beta) = \pi_\alpha(B)\, \pi_\alpha(A_i) \times \pi_\alpha(A_i)$. Conversely, given a family (π_α)

of representations of A, with the domain $\mathcal{D}_\alpha$ dense in Hilbert space H_α, one can construct $H = \oplus_{\alpha \in I} H_\alpha$ and $\pi = \oplus_\alpha \pi_\alpha$ in an obvious way.

Theorem 7.2. Let $\mathcal{B}$ be an ultraweakly closed subspace of $B(\mathcal{D}, \mathcal{D})$ satisfying condition II, and π be a normal representation of $\mathcal{B}$, with domain $\mathcal{D}_\pi$. Then, π can be identified with the direct sum $\oplus_{\alpha \in I} \pi_\alpha$ (I being some index set) of a family π_α of G.N.S. constructions $(\pi_\alpha, H_\alpha, \mathcal{D}_\alpha, \zeta_\alpha)$ associated to the positive linear form f_α on $\mathcal{B}$, for $\alpha \in I$.

More precisely, there exists a unitary operator U from the Hilbert space H_π completion of $\mathcal{D}_\pi$ onto the Hilbert space $\oplus_{\alpha \in I} H_\alpha$, such that $U(\mathcal{D})$ is exactly the set of σ-convergent families, with values in $\mathcal{D}_\alpha$.

Lemma 7.6. Take $\mathcal{B}$, $\mathcal{D}$, $\mathcal{D}_\pi$, ... of the preceding theorem. Let $\xi \in \mathcal{D}_\pi$, K be the closed linear span of $\pi(A_{id})\xi$ in H_π and f be the linear form on A given by $f(\beta) = (\pi(\beta)\xi, \xi)$, for $\beta \in A$. Then, f is normal on $\mathcal{B}$ and the G.N.S. representation $(\pi_f, H_f, \mathcal{D}_f, \zeta_f)$ associated with f coïncides with the subrepresentation of π associated with K.

This follows from lemma 7.5, and hence the theorem by Zorn's lemma.

Corollary 7.1. Let $A = \cup_i A_{A_i}$ be a $\star$-algebra satisfying condition I, and π be a representation (respectively an ultraweakly continuous representation) of A, with domain $\mathcal{D}_\pi$ in the Hilbert space H_π. Then, π can be identified with the direct sum of some family of G.N.S. representations associated to positive (respectively positive ultraweakly continuous) linear forms on A.

The proof rests on the equality $\mathcal{D}_\pi = \cap_{i \geq 0} \pi(A_i^{-1})(H)$, and is left to the reader.

Lemma 7.7. Let $\mathcal{B}$ be an ultraweakly closed subspace of $B(\mathcal{D}, \mathcal{D})$, satisfying condition II, and x be a vector in the Hilbert space H. Then, the linear form $\omega_{x,x}$ (defined on $\mathcal{B}_{id}$) extends in a positive linear form f on $\mathcal{B}$ iff x belongs to $\mathcal{D}$.

Proof. It is an obvious fact that $\omega_{x,x}$ is defined on $\mathcal{B}$ as soon as x belongs to $\mathcal{D}$, in which case the positive linear form so obtained is normal on $\mathcal{B}$. Conversely, let $x \in H$ and f be a positive extension of $\omega_{x,x}$ to the whole $\mathcal{B}$. It is sufficient to show that $x \in \mathrm{Dom}(\bar{\Delta})$, where Δ is an element of some cofinal sequence A_i in A^+. Let $\mathcal{D}_\Delta = \cap_{n\geq 0} \mathrm{Dom}(\bar{\Delta}^n)$ and p_n be a sequence, increasing to Id, of spectral projectors of $\bar{\Delta}$, such that $p_n(H) \subset \mathcal{D}_\Delta$. Then, $\mathcal{D} \subset \mathcal{D}_\Delta$, $p_n\bar{\Delta} = \bar{\Delta}p_n$ on $\mathcal{D}_\Delta$ and, putting $x_n = p_n x$, it is clear that $\|x_n - x\| \to 0$ and $x_n \in \mathrm{Dom}(\bar{\Delta})$. Moreover,

$$\|\bar{\Delta}x_n\|^2 = \|\bar{\Delta}p_n x\|^2 = \omega_{x,x}((\Delta p_n)^2) \leq f(\Delta^2)$$

shows that $\|\bar{\Delta}x_n\|$ is a bounded sequence, hence $x \in \mathrm{Dom}(\bar{\Delta})$, i.e., $x \in \mathcal{D}$, see [12] p. 315. Finally, the two positive linear forms f and $\omega_{x,x}$ are defined on $\mathcal{B}$ and coïncide on $\mathcal{B}_{id}$, hence coïncide the whole $\mathcal{B}$, since $\mathcal{B}_{id}$ is ρ-dense in $\mathcal{B}$.

Proposition 7.7. Take $\mathcal{B}$ as in the preceding lemma, and let f be a positive linear form on $\mathcal{B}$. Then, f is normal iff the restriction f_{id} of f to $\mathcal{B}_{id}$ is normal.

Proof. Let x_i be a sequence of vectors in H such that $f(T) = \Sigma_i (Tx_i, x_i)$ for all $T \in \mathcal{B}_{id}$. Let K be the Hilbert space $\ell^2(\mathbb{N})$ and $\pi: \beta \in \mathcal{B} \to \pi(\beta) = \beta \otimes 1 \in \mathcal{B} \otimes \mathbb{C}_K$ be the ampliation map. Then, formula $f_1(\beta \otimes 1) = f(\beta)$ defines a positive linear form on $\mathcal{B} \otimes \mathbb{C}_K$, and, due to $(\mathcal{B} \otimes \mathbb{C}_K)_{id} = \mathcal{B}_{id} \otimes \mathbb{C}_K$, we get $f_1(u) = \omega_{x,x}(u)$ for $u \in \mathcal{B}_{id} \otimes \mathbb{C}_K$, where $x = (x_i) \in K$. Lemma 7.7 shows that x belongs to the natural domain of $\mathcal{B} \otimes \mathbb{C}_K$ and that f_1 must coïncide with $\omega_{x,x}$ on whole $\mathcal{B} \otimes \mathbb{C}_K$, implying that f is normal on $\mathcal{B}$.

Corollary 7.2. Let $\mathcal{B}$ be as in lemma 7.7, g being a positive normal form on $\mathcal{B}_{id}$, and (π_g, H_g, ζ_g) be the G.N.S. representation of the von Neumann algebra $\mathcal{B}_{id}$. Let π be the normal lifting of π_g to $\mathcal{B}$, with domain $\mathcal{D}_\pi \subset H_g$. Then the vector ζ_g (cyclic for $\pi_g(M)$) is in $\mathcal{D}_\pi$ iff g extends in a positive linear form on $\mathcal{B}$.

Proof. If g extends to a positive linear form g_1 on $\mathcal{B}$, then g_1 is normal by proposition 7.7. Let $(\pi_1, H_1, \mathcal{D}_1, \zeta_1)$ be the G.N.S. representation of $\mathcal{B}$ associated to g_1. Then, for $T \in \mathcal{B}_{id}$,

$$g(T) = (\pi_g(T)\ \zeta_g,\ \zeta_g) = (\pi_1(T)\ \zeta_1,\ \zeta_1)$$

and lemma 7.7 leads to $\zeta_g \in \mathcal{D}_\pi$, due to $\zeta_1 \in \mathcal{D}_1$. Conversely, if $\zeta_g \in \mathcal{D}_\pi$, then the formula

$$g_1(\beta) = \pi(\beta)(\zeta_g,\ \zeta_g)$$

makes sense, and defines a positive extension of g to $\mathcal{B}$.

In fact:

Proposition 7.8. Take $\mathcal{B}$ as before, let M be the von Neumann algebra $M = \mathcal{B}_{id}$ and $P_{\mathcal{B}}$ (respectively P_M) be the predual of $\mathcal{B}$ (respectively M). The canonical map from $P_{\mathcal{B}}$ into P_M is injective continuous from the Fréchet space $P_{\mathcal{B}}$ into the Banach space P_M and has a dense range.

As an obvious consequence of this, the Fréchet space $P_{\mathcal{B}}$ is separable iff the Banach space P_M has the same property.

Proof. The canonical injection from M into $(\mathcal{B},\ \rho)$ is continuous and induces, by transposition, a continuous linear map from the Fréchet space $\mathcal{B}^\rho$ (strong dual of $(\mathcal{B},\ \rho)$) onto the Banach space M' (strong dual of M). Restricting to $P_{\mathcal{B}}$, we get a continuous linear map from $P_{\mathcal{B}}$ into P_M, injective since M is σ-weakly dense in $\mathcal{B}$. Now, $P_{\mathcal{B}}$ (respectively P_M) is the canonical image of the projective tensor product $\mathcal{D} \hat{\otimes} \mathcal{D}$ (respectively $H \hat{\otimes} H$) by the map $x \otimes y \mapsto \omega_{x,y}$, and our proposition follows from the density of $\mathcal{D} \hat{\otimes} \mathcal{D}$ into $H \hat{\otimes} H$.

We now show that G.N.S. representation may be performed for positive linear forms f on suitable spaces A satisfying condition II (with natural domain $\mathcal{D}$).

Definition 7.2. A positive linear form f on A give rise to a G.N.S. representation iff there exists a representation π of A, in sense of definition 7.1, in some Hilbert space H_π with dense domain $\mathcal{D}_\pi$ such that $f(\beta) = (\pi(\beta)\zeta,\ \zeta)$ for all $\beta \in A$, where $\zeta \in \mathcal{D}_\pi$ is a cyclic vector for $\pi(A_{id})$.

Such a representation, when it exists, is unique up to a unitary operator exchanging domain, by lemma 7.3. When any $f \geq 0$ on A give rise to a G.N.S. representation, we say for short that A admits G.N.S. representations.

Theorem 7.3. Let A be a space satisfying condition II and let $\mathcal{B} = A \cap L(\mathcal{D})$ be ρ-dense in A. Then, to each positive linear form f on A, there corresponds a unique G.N.S. representation π_f of A (up to an unitary operator exchanging domain). The representation π_f has the following properties:

1°/ the restriction of π_f to A_{id} is the G.N.S. representation of the involutive algebra A_{id}, associated to the restriction f_{id} of f to A_{id}

2°/ The G.N.S. representation of the $\star$-algebra $\mathcal{B}$, associated to the restriction of f to $\mathcal{B}$, coïncides with the restriction of π_f to $\mathcal{B}$.

As seen previously, $L(\mathcal{D})$ denotes the subset of $B(\mathcal{D}, \mathcal{D})$, consisting of operators T, such that Dom $T \supset \mathcal{D}$, Dom $T^+ \supset \mathcal{D}$, $T\,\mathcal{D} \subset \mathcal{D}$, $T^*\,\mathcal{D} \subset \mathcal{D}$. If $B \in \mathcal{B}$, then $\pi_f(B) \in L(\mathcal{D}_\pi)$ by proposition 7.10.

Proof. We first show that $\mathcal{B}$ is a $\star$-algebra containing A_i, A_i^{-1}, with natural domain $\mathcal{D}$, and satisfying condition I. Let $T_1, T_2 \in \mathcal{B}$; since $\mathcal{B} \subset L(\mathcal{D})$, there exists $i \geq 0$ and finite constants M_1, M_2, M such that, for all $x \in \mathcal{D}$,

$$\|T_1 x\| \leq M_1\|A_i x\| \leq M_1\|A_i^2 x\|$$

$$\|T_2 x\| \leq M_2\|A_i x\| \leq M_2\|A_i^2 x\|$$

$$\|T_1 T_2 x\| \leq M\|A_i^2 x\|.$$

One has $T_j = B_j A_i \times A_i$, with $B_j \in A_{id}$, $j = 1, 2$, and $T_1 T_2 = C A_i \times A_i$ with $C \in L(H)$. In order to prove that $C \in A_{id}$ we note that $B_j = T_j A_i^{-1} \times A_i^{-1}$, leading to $B_j \mathcal{D} \subset \mathcal{D}$, $B_j^*(\mathcal{D}) \subset \mathcal{D}$ and $A_i B_j = T_j A_i^{-1}$ on $\mathcal{D}$, for $j = 1, 2$. As A_{id} is an involutive algebra, $B_j A_i^{-1} \in A_{id}$ and $A_i^{-1} B_j \in A$, thus $A_i B_j = (B_j A_i^{-1}) A_i \times A_i$ and $(A_i^{-1} B_j) A_i \times A_i = B_j A_i$ belong to A. Now, $T_1 T_2 = (A_i^{-1} T_1 T_2 A_i^{-1}) A_i \times A_i$ implies that $C = A_i^{-1} T_1 T_2 A_i^{-1}$ $= (A_i (B_1)^* (A_i B_2) \in A_{id}$. Other properties of $\mathcal{B}$ are straightforward; hence, we get $\mathcal{B} = \cup_{i \geq 0}\, \mathcal{B}_{id}\, A_i \times A_i$.

Let f_0 be the restriction of f to $\mathcal{B}$, π_0 the G.N.S. representation of the $\star$-algebra $\mathcal{B}$ associated to f_0, acting in the Hilbert space H_0, with natural dense domain $\mathcal{D}_0 = \cap_{i\geq 0} \pi_0(A_i^{-1})H_0$ (due to $\pi_0(A_i)\mathcal{D}_0 = \mathcal{D}_0$) and cyclic vector $\zeta_0 \in \mathcal{D}_0 \supset \pi_0(\mathcal{B})\zeta_0$. Since $\pi_0 \geq 0$, the map π_0: $\mathcal{B} \to B(\mathcal{D}_0, \mathcal{D}_0)$ is ρ-continuous. As $(\mathcal{A}, \rho)$ induces, on the subspace $\mathcal{B}$ the given topology $(\mathcal{B}, \rho)$, π_0 has a unique continuous extension π from $(\mathcal{A}, \rho)$ into the completion $\widetilde{B}$ of $(B(\mathcal{D}_0, \mathcal{D}_0), \rho)$. Any bounded set of the DF-space $\mathcal{A}$ is contained in the closure (in $\mathcal{A}$) of a suitable bounded set of $(\mathcal{B}, \rho)$, and $B(\mathcal{D}_0, \mathcal{D}_0)$ with weak topology $\sigma(B(\mathcal{D}_0, \mathcal{D}_0), \mathcal{D}_0 \otimes \mathcal{D}_0)$ is quasi-complete ($\mathcal{D}_0 \otimes \mathcal{D}_o$ being barreled). The unique continuous extension of π_0 viewed as a map from $(\mathcal{A}, \rho)$ into $B(\mathcal{D}_0, \mathcal{D}_0)$ weakly coïncides with π, since ρ is finer than the weak topology mentioned, and satisfies $\pi(A) \subset B(\mathcal{D}_0, \mathcal{D}_0)$.

We now show that $\pi \geq 0$ on $\mathcal{A}$. From theorem 2.2, $\mathcal{B}_{id}$ is ρ-dense in $\mathcal{B}$, so that $\mathcal{B}_{id}$ is ρ-dense in $\mathcal{A}_{id}$. The maps $\beta \to \beta A_i \times A_i$ and $\beta \to \beta A_i^{-1} \times A_i^{-1}$ are homeomorphisms from $(\mathcal{A}, \rho)$ (resp. $(\mathcal{B}, \rho)$) onto itself, so that $\mathcal{B}_{A_i^2}$ is ρ-dense in $\mathcal{A}_{A_i^2}$, for every $i \geq 0$.

Let $0 \leq B \in \mathcal{A}_{id}$. Choosing a net B_α in $\mathcal{B}_{id}$ such that $B_\alpha \to B$ for ρ-topology when $\alpha \to \infty$ we get, for any $S_0 \in \mathcal{B}_{id}$,

$$(\pi_0(B_\alpha)\pi(S_0)\zeta_0, \pi(S_0)\zeta_0) = f(S_0^* B_\alpha S_0) =$$

$$= f(B_\alpha S_0 \times S_0).$$

Since π is continuous, $\pi_0(B_\alpha) \to \pi(B)$ in $(B(\mathcal{D}_{\pi_0}, \mathcal{D}_{\pi_0}), \rho)$ and, due to $\pi(S_0)\zeta_0 = u \in \mathcal{D}_\pi$, we find that $(\pi_0(B_\alpha)u, u) \to \pi(B)(u, u)$ for $u \in \pi_0(\mathcal{B}_{id})\zeta_0$. From $S_0\mathcal{D} \subset \mathcal{D}$, we find that $B_\alpha S_0 \times S_0$ tends to $BS_0 \times S_0 \geq 0$ in $(\mathcal{B}, \rho)$ and, from the continuity of f, we see that

$$\pi(B)(\pi(S_0)\zeta_0, \quad \pi(S_0)\zeta_0) = f(BS_0 \times S_0) \geq 0$$

implies that $\pi(B)(u, u) \geq 0$ for $u \in \pi(A_{id})\zeta_0$. In the same way, $B \leq \|B\|1$ gives $(\pi(\|B\|1 - B)(u, u)) \geq 0$, i.e.,

$$0 \leq \pi(B)(u, u) \leq \|B\|(u, u),$$

and lemma 7.2 shows that $\pi(B)$ (or its closure) must be a bounded operator in the Hilbert space H_π. It follows directly that $\pi(ST) = \pi(S)\pi(T)$, $\pi(S^*) = \pi(S)^*$ for S, T in A_{id}. Now, $\beta = BA_i \times A_i$ is the ρ-limit of the net $B_\alpha A_i \times A_i = A_i B_\alpha A_i \in B$, and the formula

$$(\pi(B_\alpha)\pi_0(A_i)\pi_0(S_0)\zeta_0, \ \pi_0(A_i)\pi(S_0)\zeta_0) = f(S_0^* A_i B_\alpha A_i S_0)$$

implies, for $u \in \pi(B_{id})\zeta_0$, that

$$(\pi(B)\pi(A_i)u, \ \pi(A_i)u) \geq 0,$$

since $B_\alpha A_i S \times A_i S$ tends to $BA_i S \times A_i S \geq 0$ in (A, ρ), due to $(A_i S)\mathcal{D} \subset \mathcal{D}$. Arguing as above, we find, for $u \in \pi(A_{id})\zeta_0$, that

$$(\pi(B)\pi(A_i)u, \ \pi(A_i)u) \leq \|B\| (\pi(A_i^2)u, \ u),$$

showing that $\pi(B)\pi(A_i) \times \pi(A_i)$ defined on $\pi(A_{id})\zeta_0 \times \pi(A_{id})\zeta_0$ has a unique continuous extension to Dom $\overline{\pi(A_i)}\times$Dom $\overline{\pi(A_i)}$ and that $\pi(\beta) = \pi(B)\pi(A_i) \times \pi(A_i) \geq 0$ on $\mathcal{D}_\pi \times \mathcal{D}_\pi$. Finally, ζ_0 being cyclic for $\pi_0(B_{id})$, it is also cyclic for $\pi(A_{id})$ and, by standard arguments, we see that (up to a unitary operator) the usual G.N.S. representation of the algebra A_{id} may be identified to the restriction of π to A_{id}; thus $\pi = \pi_f$ has all the properties required. The question of unicity is straightforward, since any positive map π on A, being continuous, is completely determined by its restriction to B.

Lemma 7.8. Let $A = \cup_{n \geq 0} A_{\Delta^n}$ be a space satisfying condition II with natural domain $\mathcal{D}$ (with $\Delta \geq 1$, $\Delta\mathcal{D} = \mathcal{D}$). If the von Neumann algebra P generated by Δ^{-1} is contained in A_{id}, then $A_{id} \cap L(\mathcal{D})$ is ρ-dense in A.

Proof. The natural domain of the $*$-algebra B generated by P, Δ, Δ^{-1} is exactly $\mathcal{D}$ and, from theorem 1.1, B is an ultraweakly closed $*$-algebra satisfying condition I. Let $\beta \in A$, and $B \in A_{id}$, $j \in \mathbb{N}$, such that $\beta = B\Delta^j \times \Delta^j$; for simplicity, we choose $j = 1$. By lemma 2.4, there exists a sequence B_n of elements of B_{id} satisfying $B_n(H) \subset \mathcal{D}$, $0 \leq B_n \leq \Delta$, a sequence $\varepsilon(n)$ of positive reals tending to zero and an integer $p \geq 0$, such that one has $\|(\Delta - B_n)x\| \leq \varepsilon(n)\|\Delta^p x\|$ for all $x \in \mathcal{D}$. One has $B_n B B_n \in A_{id} \cap L(\mathcal{D})$ and, for x in $\mathcal{D}$,

$$|(B\Delta x,\ \Delta x) - (B\ B_n x,\ B_n x)|$$

$$\leq |(B\Delta x,\ \Delta x) - (B\Delta x,\ \Delta_n x)| + |(B\Delta x,\ \Delta_n x) - (B\ B_n x,\ B_n x)|$$

$$\leq \|B\|\ \|\Delta x\|\ \|\Delta x - B_n x\| + \|B\|\ \|\Delta x - B_n x\|\ \|B_n x\|$$

$$\leq \varepsilon(n)\|B\|(\|\Delta x\|\ \|\Delta^P x\| + \|\Delta^P x\|\ \|B_n x\|$$

$$\leq 2\varepsilon(n)\|B\|(\Delta^{2p+2}x,\ x)$$

thus β is the limit in the normed space $(A_{\Delta^{2P+2}}, \rho_{\Delta^{2P+2}})$ of a sequence of elements of $A_{id} \cap L(\mathcal{D})$. From the continuity of the imbedding of $A_{\Delta^{2P+2}}$ into (A, ρ), we get the lemma.

Theorem 7.4. Let A be an ultraweakly closed space with cofinal abelian sequence A_i satisfying condition II. Then, every positive linear form g on A gives rise to a G.N.S. representation π_g of A.

Proof. For every integer $j \geq 0$, we put $\mathcal{D}_j = \cap_{k\geq 0} A_j^{-k}(H)$, $A_j = \cup_{k\geq 0}\ M\ A_j^k \times A_j^k$, where M is the von Neumann algebra A_{id}; clearly, A_j is an ultraweakly closed space of $B(\mathcal{D}_j, \mathcal{D}_j)$ satisfying condition II. The ρ-topology of A_j, which refers to $\mathcal{D}_j \times \mathcal{D}_j$, will be denoted ρ_j, and π is the G.N.S. representation of the involutive algebra A_{id} associated to the restriction f_{id} of f to A_{id}. acting in the Hilbert space H_π with cyclic vector ζ. By lemma 7.8 and theorem 7.3, π_{id} has a unique ρ_j-continuous extension π_j from (A_j, ρ_j) into $B(\mathcal{D}_{\pi_j}, \mathcal{D}_{\pi_j})$, where $\mathcal{D}_{\pi_j}$ is the Fréchet space

$\mathcal{D}_{\pi_j} = \cap_{k\geq 0}\ \pi_{id}\ (A_j^{-k})(H_\pi)$ and π_j clearly coïncide on $A_j \cap L(\mathcal{D}_j)$ with the G.N.S. representation associated to the restriction f_j of f to A_j ($\mathcal{D}$ is dense in $\mathcal{D}_j$ by proposition 3.5). Let P_π be the abelian von Neumann algebra generated by all $\pi_j(A_j^{-k}) = \pi_{id}(A_j^{-k})$, for $k \geq 0$, $j \geq 0$. Given $j \geq 0$, one can find a sequence E_n of projectors belonging to the von Neumann algebra P_j generated by A_j^{-1}, with $E_n(H) \subset \mathcal{D}_j$ for all $n \geq 0$, and E_n converging to Id in (A_j, ρ_j), by lemma 2.4. By the closed graph theorem, $A_j^k E_n$ are bounded operators in P_j, for $k \geq 0$, therefore $\pi_j(A_j^k E_n) = \pi_j(A_j)^k \pi_j(E_n)$ are bounded operators in $L(H_\pi)$, implying that $\pi_j(E_n)(H_\pi) \subset \pi_j(A_j)^{-k}(H_\pi)$ for all $k \geq 0$, i.e., $\pi_j(E_n)H_\pi \subset \mathcal{D}_{\pi_j}$; it follows that $\mathcal{D}_{\pi_j}$ is an essentially dense domain, relative to P_π.

Introducing $\mathcal{D}_\pi = \cap_{j\geq 0} \mathcal{D}_{\pi_j}$, we get (by [17]) that $\mathcal{D}_\pi$ is dense in the Hilbert space H_π. Clearly, $\pi(A_j^{-1})\pi(A_k^{-1}) = \pi(A_j^{-1})\pi(A_k^{-1})$, for $j \geq 0$, $k \geq 0$, implies that $\pi(A_j^{-1})\, \mathcal{D}_{\pi_k} \subset \mathcal{D}_{\pi_k}$, i.e., $\pi(A_j^{-1})\mathcal{D}_\pi \subset \mathcal{D}_\pi$. Let Δ be any power of some A_k; given v in $\mathcal{D}_\pi$, one has $A_j\ \Delta \in A$, hence $A_j\Delta \leq MA_i$ for some $i \geq 0$ and $M < +\infty$. Therefore, $M^{-1}\pi(A_i^{-1}) \leq \pi(A_j^{-1})\pi(\Delta^{-1})$ implies

$$\text{Dom}\ \overline{\pi_i(A_i)} = \pi(A_i^{-1})(H_\pi) \subset \pi(A_j^{-1}\Delta^{-1})(H_\pi) = \text{Dom}\ \pi_i(A_j\Delta)$$

and $v \in \pi(A_i^{-1})H_\pi$ leads to $v = \pi(A_j^{-1})\pi(\Delta^{-1})u$ with $u \in H_\pi$, i.e., $v \in \text{Dom}\ \overline{\pi_j(A_j)}$ and $\pi_j(A_j)v \in \mathcal{D}_\pi$, since this holds for any Δ. Finally, $\pi_j(A_j)\mathcal{D}_\pi = \mathcal{D}_\pi$.
By proposition 3.5, each $\mathcal{D}_{\pi_j}$ is dense in the Fréchet space $\mathcal{D}_\pi$: thus, all representations π_i, $i \geq 0$, can be viewed as representations with values in $B(\mathcal{D}_\pi, \mathcal{D}_\pi)$ with its natural ρ-topology, leading to a representation $\pi^\cdot$ of A whose restriction to each A_j coincides with π_j.

Remark 7.2. In order to complete theorem 7.3 (keeping its notations), we give an alternative description of the G.N.S. representation associated to a positive form f on A. Since A_0 is ρ-dense in $A = \cup_{i\geq 0} AA_i$ (A_0 = Id), and (A, ρ) induces (A_0, ρ) by theorem 1 [1], π_f is exactly the unique ρ-continuous linear extension to A of the classical G.N.S. representation π_{f_0} of the $\star$-algebra A_0 associated to the positive linear form f_0, restriction of the given f to A_0 ($\pi_{f_0}(A_0)$ with domain $\mathcal{D}_f = \mathcal{D}_{f_0} = \cap_{i\geq 0}\ \pi_{f_0}((A_i^{-1})H_{f_0})$. For every integer $i \geq 0$, A_0 is endowed with a scalar product $< , >_i$, where

$$<S, T>_i = f(T^* A_i^2 S), \quad S, T \in A_0,$$

is an a priori non-separated pre-Hilbert space, since

$$|\langle S, T\rangle|_i = |f((A_iT)^*A_iS)| \le f((A_iS)^*A_iS)^{\frac{1}{2}} f(A_iT)^*A_iT)^{\frac{1}{2}}$$

$$\le \langle S, S\rangle_i^{\frac{1}{2}} \langle T, T\rangle_i^{\frac{1}{2}}.$$

The linear subsets $N_i = \{T \in \mathcal{A}_0 | \langle T, T\rangle_i = 0\}$ do not depend on $i \ge 0$ and coïncide with $N_0 = \{T \in \mathcal{A}_0 | \langle T, T\rangle_0 = f(T^*T) = 0\}$. Indeed, as $A_i^2 \ge \mathrm{Id}$, the formula

$$f(T^*T) \le f(T^*A_i^2T)$$

gives $N_i \subset N_0$ and, for $T \in N_0$,

$$f(T^*A_i^2T) \le f(T^*T)^{\frac{1}{2}} f(S^*S)^{\frac{1}{2}},$$

where $S = A_i^2T$, gives $T \in N_i$, hence $N_i = N_0$ for $i \ge 0$. Let $j : \mathcal{A}_0 \to \mathcal{A}_0/N_0$ be the natural quotient map. By the definition of the G.N.S. representation π_{f_0}, $\mathcal{D}_{f_0}$ is the Fréchet space completion of $\mathcal{A}_0/N_0$ endowed with the sequence of semi-norms

$$j(T) \in \mathcal{A}_0/N_0 \mapsto \|\pi_{f_0}(A_i)j(T)\| = f(T^*A_i^2T)^{\frac{1}{2}},$$

and π_{f_0} is the quotient in $\mathcal{A}_0/N_0$ of the left regular representation $T \to \pi(T)$ of $\mathcal{A}_0$, with $\pi(T)S = TS$.

Let us assume f_0 faithful, i.e., the conditions $0 \le T \in \mathcal{A}_0$ and $f(T) = 0$ imply that $T = 0$. For every $i \ge 0$, let f_i be the linear positive form on $\mathcal{A}_0$ (or on $\mathcal{A}$ using proposition 2.3 1°/) as defined by $f_i(T) = f(A_iTA_i)$ for $T \in \mathcal{A}_0$. For $S, T \in \mathcal{A}_0$ we put $_i\langle S, T\rangle = f_i(T^*S) = f(A_iT^*SA_i)$. Due to

$$|f_i(S^*T)| = |f((SA_i)^*TA_i)| \le f(A_iT^*TA_i)^{\frac{1}{2}} f(A_iS^*SA_i)^{\frac{1}{2}},$$

$\mathcal{A}_0$ is a separated pre-Hilbert space for $_i\langle\ ,\ \rangle$, since conditions $_i\langle S, S\rangle = 0$ give $f((S A_i)^*SA_i) = 0$ i.e., $((S^*A_i)^*SA_i\varphi, \varphi) = 0$ for all $\varphi \in \mathcal{D}$, implying that $\|SA_i\varphi\| = 0$, hence $S = SA_i\, A_i^{-1} = 0A_i^{-1} = 0$. The sequence of semi-norms

$S \in A_0 \mapsto f(A_i S^* S A_i)^{\frac{1}{2}}$ makes A_0 a new metric topological space, in general distinct from A_0, with the semi-norms $S \in A \mapsto f(S^* A_i^2 S)^{\frac{1}{2}}$.

We shall see in paragraph 10 that this new topology corresponds to a choice of a privilegiate commutant of A_0 (see proposition 10.2).

Proposition 7.9. Let A be a space with condition II, and π be a representation (of A) acting in some Hilbert space H_π with domain $\mathcal{D}_\pi$. The von Neumann algebras generated by $\pi(A_{id})$ and $\pi(A)_{id}$ coïncide. If $\zeta_\pi \in \mathcal{D}_\pi$ is a cyclic vector for $\pi(A_{id})$, then $\pi(A_{id})\zeta_\pi$ is dense in the Fréchet space $\mathcal{D}_\pi$ as soon as it is contained in $\mathcal{D}_\pi$.

Proof. Due to $\pi(A_{id}) \subset \pi(A)_{id}$, it suffices to show that $\pi(A_{id})'' \supset [\pi(A)_{id}]''$. Let $T \in L(H_\pi)$ which commutes with $\pi(A_{id})$. From $\pi(A_i)\mathcal{D}_\pi = \mathcal{D}_\pi$, and $\pi(A_i^{-1}) \in \pi(A_{id})$, we get $T\mathcal{D}_\pi \subset \mathcal{D}_\pi$ $T^*\mathcal{D}_\pi \subset \mathcal{D}_\pi$ and $T\pi(A_i) = \pi(A_i)T$ on $\mathcal{D}_\pi$ for $i \geq 0$. Let $\beta = BA_i x A_i$ with $B \in A_{id}$ such that $\pi(\beta) \in \pi(A)_{id}$. One has for $x, y \in \mathcal{D}_\pi$,

$$\begin{aligned}(\pi(\beta)Tx, y) &= (\pi(B)\pi(A_i)Tx, \pi(A_i)y)\\ &= (\pi(B)T\pi(A_i)x, \pi(A_i)y)\\ &= (T\pi(B)\pi(A_i)x, \pi(A_i)y)\end{aligned}$$

and

$$\begin{aligned}(T\pi(\beta)x, y) &= (\pi(\beta)x, T^*y) = (\pi(B)\pi(A_i)x, \pi(A_i)T^*y)\\ &= (\pi(B)\pi(A_i)x, T^*\pi(A_i)y),\end{aligned}$$

hence $\pi(\beta)$ commutes with T, showing the first part of the proposition. Now, it follows that the closures in H_π of the orbit of ζ_π under $\pi(A)_{id}$ or under $[\pi(A)]_{id}$ coïncide, thus, by the Hahn-Banach theorem, a continuous linear form f on $\mathcal{D}_\pi$, vanishing on $\pi(A_{id})\zeta_\pi$, must be zero, as can be seen by slightly modifying the proof of proposition 3.11.

Theorem 7.5. Let A be a space with condition II and cofinal abelian sequence A_i. To each positive linear form g on A corresponds a (unique) G.N.S. representation π_g of A.

Proof. We need to prove the existence of π_g, as seen in definition 7.2. Let $\mathcal{B} = B(\mathcal{D}, \mathcal{D})$ and $f \geq 0$ on $\mathcal{B}$ be a positive linear extension of g. As $\mathcal{B}$ is ultraweakly closed, there exists (by theorem 7.4) a representation π_f of $\mathcal{B}$, with cyclic vector $\zeta_f \in \mathcal{D}_f$ acting in some Hilbert space H_f, where $\mathcal{D}_f \equiv \cap_{i\geq 0} \pi_f (A_i^{-1})H_f$. Looking at the restriction of π_f to A_{id}, we introduce the closure H_0 in the Hilbert space H_f of $\pi_f(A_{id})\zeta_f$, and let E be the corresponding projector. Clearly, E is in the von Neumann algebra $\pi_f(A_{id})'$ (therefore it commutes with all $\pi_f(A_i^{-1})$), hence $E\,\mathcal{D}_f \subset \mathcal{D}_f$. Put $\mathcal{D}_g = E\mathcal{D}_f$ with its natural topology (see lemma 7.5). We note here that $\pi_f(A)$ with domain $\mathcal{D}_f$ does not necessarily satisfy condition II, however its σ-weak closure relative to $\mathcal{D}_f \hat{\otimes} \mathcal{D}_f$ satisfies condition II, due to proposition 7.9. For $x \in A$, let $\pi_g(x)$ be the restriction of $\pi_f(x) \in B(\mathcal{D}_f, \mathcal{D}_f)$ to $\mathcal{D}_g \times \mathcal{D}_g$, i.e., $\pi_g(x) \in B(\mathcal{D}_g, \mathcal{D}_g)$ as in theorem 4.1. It is now straightforward that $x \in A \mapsto \pi_g(x) \in B(\mathcal{D}_g, \mathcal{D}_g)$ is a representation in the Hilbert space $H_g = E(H_f)$ and $\zeta_g = E\zeta_f \in \mathcal{D}_g$, being a cyclic vector in H_g for $\pi(A_{id})_E$, i.e., for $\pi_g(A_{id})$, we find, from $\pi_f(A_i)E\mathcal{D}_f \subset E\mathcal{D}_f$ and $\pi_f(A_i^{-1})E\mathcal{D}_f \subset E\mathcal{D}_f$ that $\pi_g(A_i)\mathcal{D}_g = \mathcal{D}_g$, hence π_g is the G.N.S. representation we are looking for.

Remark 7.3. In practice, it suffices to perform the G.N.S. representation for a space A with condition II of the form $A = \cup_{n\geq 0} A_{\Delta^n}$, $\Delta \geq \mathrm{Id}$, $\Delta\mathcal{D} = \mathcal{D}$, and we recall that this point of view has already be mentioned in remark 1.1. Indeed, an $f \geq 0$ on a space A with condition II and cofinal sequence A_i - not necessarily abelian - induces on each $A_i = \cup_{n\geq 0} A_{id}A_i^n \times A_i^n$ a positive linear form f_i, for which theorem 7.5 applies, and one may consider the sequence of representation π_{f_i} so obtained (all identical on A_{id}) as the G.N.S. representation of (A, f). For an ultraweakly closed A and $f \geq 0$ normal, this too large generality is avoided, since the G.N.S. representation of (A, f) is simply the normal lifting to A (see proposition 7.4) of the G.N.S. representation of the couple (M, f) where $M = A_{id}$.

2°/ Let $f \geq 0$ on A and π_f be the G.N.S. representation of A. Let $A_0 = A \cap L(\mathcal{D})$ be ρ-dense in A and

$j : S \in \mathcal{A}_0 \mapsto j(S) = \pi_f(S)\zeta_f \in \mathcal{D}_f$ be the usual map.
One has, for $\beta = BA_i \times A_i \in \mathcal{A}$ with $B \in \mathcal{A}_{id}$,

$$f(BA_iS \times A_iT) = \pi(\beta)(j(S),\ j(T))$$
$$= (\pi(B)j(A_iS),\ j(A_iT))$$

for all $S, T \in \mathcal{A}_0$. This may be seen as follows. Put, for every $S, T \in \mathcal{A}_0$

$$(S, T)_\beta = f(\beta(S\cdot,\ T\cdot)).$$

One may assume $\beta = \beta^*$, and we choose $M < +\infty$ such that $|\beta| \leq MA_i^2$. Therefore, $0 \leq \beta + MA_i^2 \leq 2MA_i^2$ leads to

$$|f((\beta + MA_i^2)(S\ \cdot,\ T\ \cdot))|$$
$$\leq f((\beta + MA_i^2)(S\ \cdot,\ S\ \cdot))^{\frac{1}{2}}\ f(\beta + MA_i^2)(T\ \cdot,\ T\ \cdot))^{\frac{1}{2}}$$

and, using

$$|(S, T)|_\beta \leq |(S, T)|_{\beta+MA_i^2} + |(S, T)|_{MA_i^2},$$

we find that

$$|(S, T)_\beta| \leq \|S\|_{MA_i^2}\|T\|_{MA_i^2} + \|S\|_{\beta+MA_i^2}\|T\|_{\beta+MA_i^2}$$
$$\leq 5\ M^2\|S\|_{A_i^2}\|T\|_{A_i^2}\ .$$

Now, one has

$$\|S\|_{A_i^2} = f(A_iS \times A_iS) = (\pi_f(A_i)j(S),\ \pi_f(A_i)j(S)),$$

hence we find a bounded $b = b^*$ in $L(H_f)$, such that

$$(S, T)_\beta = (b\pi_f(A_i)j(S),\ \pi_f(A_i)j(T)).$$

From the definition of a representation, we get $b = \pi(B)$ and the first formula follows. The second formula comes from A_iS and $A_iT \in \mathcal{A}_0$ with B in place of β.

Proposition 7.10. Take $\mathcal{A}$ with condition II and natural domain $\mathcal{D}$, and let π be a representation of $\mathcal{D}$ with domain $\mathcal{D}_\pi$ dense in some Hilbert space H_π.

1°/ For an operator $T \in A$, such that $T\mathcal{D} \subset \mathcal{D}$ and $T^*\mathcal{D} \subset \mathcal{D}$, one has $\pi(T)\mathcal{D}_\pi \subset \mathcal{D}_\pi$ and $\pi(T)^*\mathcal{D}_\pi \subset \mathcal{D}_\pi$

2°/ For an operator $T \in A$, such that $T(H) \subset \mathcal{D}$ and $T^*(H) \subset \mathcal{D}$, one has $\pi(T)H_\pi \subset \mathcal{D}_\pi$ and $\pi(T)^*H_\pi \subset \mathcal{D}_\pi$

Proof. For the first assertion, consider the $*$-algebra C generated by all A_i, A_i^{-1}, and T, and use proposition 7.3 for the restriction of π to C. For the second assertion, note that $T(H) \subset \mathcal{D}$ is equivalent to $A_i T$ bounded operator in H for all $i \geq 0$: this property holds for $\pi(A_i T) = \pi(A_i)\pi(T)$ leading to $\pi(T) \subset \mathrm{Dom}\, \pi(A_i) = \pi(A_i^{-1})H_\pi$ for all $i \geq 0$, i.e., $\pi(T) \subset \mathcal{D}_\pi$

Proposition 7.11. Let P be an abelian ultraweakly closed $*$-algebra with condition II, and π be a normal representation of P. If every character on P is normal, the same property holds for $\pi(P)$.

This proposition concerns the following situation. Let A be an ultraweakly closed space with condition II and cofinal abelian sequence A_i. Let π be a normal representation of A and P be the abelian ultraweakly closed $*$-algebra generated by all A_i, A_i^{-1}. When characters on P are normal, we see (using the remark of theorem 5.3) that H splits into a direct Hilbert sum and $\pi(A_i)$ is a direct sum of homotheties in this decomposition.

When $\mathcal{D}$ is a Schwartz space, positive linear forms on A (hence characters on P) are normal (by [10]).

For a general ultraweakly closed abelian $*$-algebra with condition II, it is easily seen that there exists a unique projector E in P, such that all characters on P_E are normal and all characters on P_{1-E} are not normal.

Proof of the proposition. As π is normal, there exists a projector $F \in P$, such that the restriction $\pi|P_F$ is an isomorphism from P_F onto $\pi(P)$ and $\pi|P_{1-F} = 0$. Characters χ on P_F are characters χ on P, such that $\chi(F) \neq 0$, hence they are normal.

Proposition 7.12. Let $A = \cup_{n\geq 0} A_{\Delta^n}$ be ultraweakly closed with condition II, $f \geq 0$ on A and π_f be the corresponding G.N.S. representation of A. The map

$$j : S \in A\cap L(\mathcal{D}) \mapsto j(S) = \pi(S)\zeta_f \in \mathcal{D}_f$$

extends continuously from (A, ρ) into the strong antidual $(\mathcal{D}_f)'$ of $\mathcal{D}_f$, i.e., for $\beta \in A$, $j(\beta)$ is a distribution on $\mathcal{D}_f$. The multiplication $\pi(S)j(T) = j(ST)$ extends continuously from $(\mathcal{D}_f)'$ strong into itself. One has, for $S, A \in L(\mathcal{D}) \cap A$,

$$<j(\beta), j(S)> = (\pi(\beta)\zeta_f, j(S)) = f(\beta(\cdot , S \cdot));$$

$$<\pi(S)j(\beta), j(A)> = <j(\beta), j(S^*A)> = (\pi(\beta)\zeta_f, j(S^*A)).$$

The antidual $\mathcal{D}_f'$ of $\mathcal{D}_f$ is the space of antilinear continuous linear forms on $\mathcal{D}_f$. We write $<f, v> = f(v)$ for the duality $<\mathcal{D}_f', \mathcal{D}_f>$. From proposition 3.4, the Hilbert space H_f is embedded linearly into $\mathcal{D}_f'$; to each $y \in H$, it corresponds the continuous antilinear form $v \in \mathcal{D}_f \mapsto <y, v> = (y, v)$.

Remark. Proposition 7.12 remains true if A is not ultraweakly closed and $A \cap L(\mathcal{D})$ is ρ-dense in A. It also remains true if f is positive normal on a general ultraweakly closed A, i.e., the cofinal sequence A_i is not necessarily abelian.

Proof. We denote by A_i any power of the Δ operator. Let $\beta = BA_i \times A_i = \beta^* \in A$, with $B \in A_{id}$ and $\bar{A}_i = \int_1^\infty \lambda \, dp_\lambda$. One has $\beta = \rho\text{-} \lim \beta_n$ with $\beta_n = BA_i^{(n)} \times A_i^{(n)}$, where $A_i^{(n)} = \int_1^n \lambda \, dp_\lambda$. By proposition 2.3, one has $\rho\text{-lim } \beta_n(\cdot, S \cdot) = \beta(\cdot, S \cdot)$, implying that $f(\beta_n(\cdot, S \cdot)) \to f(\beta(\cdot , S \cdot)$ for $S \in A \cap L(\mathcal{D})$. One has

$$f(\beta_n(\cdot , S \cdot)) = f(\beta_n \times S) = (\pi(\beta_n)\zeta_f, \pi(S)\zeta_f),$$

since $\beta_n \equiv A_i^{(n)} B A_i^{(n)}$ is bounded as an operator. Taking the limit as $n \to \infty$, we get an antilinear map defined on $j(A)$ - denoted by $j(\beta)$ - by the formula

$$<j(\beta), \pi(S)\zeta_f> = f(\beta(\cdot , S \cdot)).$$

It remains to show that $j(\beta)$ extends as a continuous form $j(\beta)$ on the Fréchet space $\mathcal{D}_f$. Putting, for $n \geq 0$, $B_n = E_n B E_n$, where $E_n = \int_1^n dp_\lambda$, we find that

$$|f(\beta_n \times S)| = |f(B_n A_i \times A_i S)|$$

$$= |(\pi(B_n)\pi(A_i)\zeta_f, \pi(A_i)\pi(S)\zeta_f)|$$

$$\leq \|\pi(B_n)\| \ \|\pi(A_i)\zeta_f\| \ \|\pi(A_i)\pi(S)\zeta_f\|$$

$$\leq \|B\| \cdot M \cdot \| \pi(A_i)j(S)\|,$$

where $M = \|\pi(A_i)\zeta_f\|$, i.e.,

$$|\langle j(\beta), j(S)\rangle| \leq c^{te}\|\pi(A_i)j(S)\|$$

is the continuity mentioned. From $f(BA_iT \times A_iS) = (\pi(\beta)j(T), j(S)$, for $S, T \in L(\mathcal{D}) \cap A$, we find that $\langle j(\beta), j(S)\rangle =$ $= (\pi(\beta)\zeta_f, j(S))$. Let β_n be introduced, with $\beta = \rho\text{-lim}\ \beta_n$. One has $\pi(S)j(\beta_n) = j(S\beta_n)$. Thus, for all $A \in A \cap L(\mathcal{D})$,

$$\langle\pi(S)j(\beta_n), j(A)\rangle = (j(S\beta_n), j(A))$$

$$= (\pi(S)\pi(\beta_n)\zeta_f, \pi(A)\zeta_f) = (\pi(\beta_n)\zeta_f, \pi(S)^*\pi(A)\zeta_f)$$

$$= (\pi(\beta_n)\zeta_f, \pi(S^*A)\zeta_f),$$

thus it tends, as $n \to \infty$, to $\langle j(\beta), j(S^*A)\rangle$. The rest of the details are clear.

APPENDIX: THE SECOND DUAL

The aim of this appendix is to describe the second dual A'' of objects on which G.N.S. representations may be performed. We are thus concerned with $*$-algebras A satisfying condition I, or with spaces A satisfying condition II and admitting a cofinal abelian sequence.

In order to lighten these notes, propositions and proofs have been written only for $*$-algebras. Similar results holds for spaces A just discussed, and the associated proofs are easy adaptations of process developed in theorems 7.3, 7.4 - due to the intervention of $L(\mathcal{D}) \cap A$. Let A be a $*$-algebra (with or without condition I) with natural domain $\mathcal{D}$. By definition, the universal representation π of A is the direct sum $\oplus_{f\geq 0}\ \pi_f$ of all G.N.S. representations π_f associated to positive linear forms f on A. Let $\mathcal{D}_\pi$ be the natural domain of $\pi(A)$, and $\mathcal{B}$ be the ultraweak closure of $\pi(A)$ in $B(\mathcal{D}_\pi, \mathcal{D}_\pi)$. As usual, A, $\pi(A)$, $\mathcal{B}$ are endowed with ρ-topology and

and A^ρ, $\pi(A)^\rho$, $\mathcal{B}^\rho$ denote the (Fréchet spaces) strong duals of these spaces. Finally, $P_\mathcal{B}$ is the predual of $\mathcal{B}$ and j is the canonical injection from $(\pi(A), \rho)$ into $(\mathcal{B}, \rho)$. It is clear that π is an isomorphism from (A, ρ) onto $(\pi(A), \rho)$ preserving ρ-norms.

The usefulness of condition I becomes apparent in

<u>Proposition 7.13</u>. The transposed map ${}^t(j \circ \pi)$ of $j \circ \pi$ restricted to the Fréchet space $P_\mathcal{B}$ is bijective from $P_\mathcal{B}$ onto the linear subset M of A' generated by positive linear forms on A. This map is an homeomorphism from the Fréchet space $P_\mathcal{B}$ onto M (endowed with topology induced by A^ρ) iff every bounded subset of $(\mathcal{B}, \rho)$ is contained in the ultraweak closure, relative to $\mathcal{D} \hat{\otimes} \mathcal{D}$ of a bounded subset of $(\pi(A), \rho)$.

<u>Proof</u>. Any ultraweakly continuous linear form on $\mathcal{B}$ is completely determined by its restriction to $\pi(A)$, thus first assertion follows from [1] theorem 2. Now, let f in the strong closure of M and f_n be a sequence in M with limit f. Quantities $(f_n - f_m)([-A_i, A_i]_A)$ tend to zero as n, $m \to \infty$, for a fixed integer i (and A_i obviously cofinal in A^+). Functions $f_n \circ \pi^{-1}$ are ultraweakly continuous on $\mathcal{B}$, so that the preceding quantities agree with $((f_n - f_m) \circ \pi)(\overline{[-\pi(A_i), \pi(A_i]}_{\pi(\mathcal{B})})$, where the bar stands for an ultraweak closure (proposition 2.2). From our assumption relating to bounded sets, we find that f_n is a Cauchy sequence in strong $\mathcal{B}^\rho$, so defining an element of $P_\mathcal{B}$ whose restriction to A clearly coïncides with f. Now, a continuous bijection between Fréchet spaces must be a topological isomorphism. Conversely, M closed in A^ρ means that $P_\mathcal{B}$ endowed with topology of uniform convergence on intervals $\overline{[-\pi(A_i), \pi(A_i]}_{\pi(A)}$, $i \in \mathbb{N}$, is a Fréchet space. Being already a Fréchet space for uniform convergence on bounded sets of $(\mathcal{B}, \rho)$, we find the coïncidence of these σ-topologies, hence the proposition.

<u>Lemma 7.9</u>. Let A be a $*$-algebra with condition I and $\mathcal{B}$ its σ-weak closure. Let us assume that ρ-continuous linear form on A are σ-weakly continuous, thus admitting a unique σ-weakly continuous extension $\tilde{f}$ to $\mathcal{B}$.

1°/ The map $f \to \tilde{f}$ is an isomorphism from the Fréchet space A^ρ onto the Fréchet space $P_\mathcal{B}$. This isomorphism preserves the natural semi-norms of these Fréchet spaces.

2°/ For T in $\mathcal{B}$, let $\hat{T}$ be the linear form on A^ρ defined by $\hat{T}(f) = \tilde{f}(T)$. The correspondence $T \to \hat{T}$ is a bijective linear map from $\mathcal{B}$ onto the second dual A'', whose restriction to A is the canonical injection from A into A''. This bijection is a topological isomorphism when $\mathcal{B}$ has the topology of the strong dual of $P_{\mathcal{B}}$ and A" the topology of the strong dual of A^{ρ}.

Proof. Let i be the continuous injection from (A, ρ) into $(\mathcal{B}, \rho)$. Transposition gives a continuous linear map $t_i : \mathcal{B}^{\rho} \to A^{\rho}$ which, restricted to $P_{\mathcal{B}}$, leads to a map $j : P_{\mathcal{B}} \to A^{\rho}$, which is bijective from our assumptions; we therefore get a topological isomorphism between these Fréchet spaces. The topology of $P_{\mathcal{B}}$ (resp. A^{ρ}) is the topology of uniform convergence on intervals $[-A_i, A_i]_{\mathcal{B}}$ (respectively $[-A_i, A_i]_A$), with $i \in \mathbb{N}$. From proposition 2.2 and the second commutant theorem, the formula

$$\sup_{T\in[-A_i,A_i]_{\mathcal{B}}} |\tilde{f}(T)| = \sup_{T\in[-A_i,A_i]_A} |f(T)|$$

holds for f in A^{ρ}, showing 1°/. For 2°/, let $T \in \mathcal{B}$, and $i \geq 0$, $M < +\infty$, such that $T \in M[-A_i, A_i]$. Semi-norms of A^{ρ} are the maps

$$f \in A^{\rho} \to \|f\|_{A_i} = \sup\{|f(S)|; \quad |S| \leq A_i, \ S \in \mathcal{B}\}$$

and the formula

$$|\hat{T}(f)| = |f(T)| \leq M\|f\|_{A_i} ,$$

for f in A^{ρ}, shows that $\hat{T} \in A"$. Conversely, any continuous linear form g on A^{ρ} may be identified to a continuous linear form on $P_{\mathcal{B}}$, due to 1°/ and, hence, corresponds to an element of $\mathcal{B}$ by the bipolar theorem. Finally, $\hat{T}(\tilde{f}) = f(T)$, for T in A, shows that $T \to \hat{T}$ restricted to A is the natural injection from A into A". A direct transposition of j gives the last assertion.

Remark 7.4. The strong dual $\tilde{\mathcal{B}}$ of the Fréchet space $P_{\mathcal{B}}$ is a complete DF space, equal as vector space to $\mathcal{B}$ = A", and $\tilde{\mathcal{B}}^+$ is normal (using theorem 1 [10]). Bounded subsets of $(\mathcal{B}, \rho)$ and $\tilde{\mathcal{B}}$ coïncide, although $(\mathcal{B}, \rho)$ is finer than $\tilde{\mathcal{B}}$. Thus, A" strong is bornological iff $\tilde{\mathcal{B}} = (\mathcal{B}, \rho)$. When bounded subsets of $\mathcal{B}$ are metrisable (such a property is true for a quasi-normable domain $\mathcal{D}$ or, more generally, for a space A satisfying the strict condition of Mackey convergence [10]), it follows from a theorem attributed to Grothendieck - see [7] - that

A" is bornological. Separability of $(\mathcal{B}, \rho)$, or of $\widetilde{\mathcal{B}}$, is also a sufficient condition.

Proposition 7.14. Let A be a $*$-algebra with condition I, π the universal representation of A, and $\mathcal{B}$ the σ-weak closure of $\pi(A)$. Then

(i) every positive form on $\mathcal{B}$ is a form $\omega_{\xi,\xi}$, with $\xi \in \mathcal{D}_\pi$. Ultraweakly continuous linear forms on $\mathcal{B}$ are weakly continuous.

(ii) Given f in A^ρ, then exists a unique, weakly continuous linear form $\tilde{f}$ on $\mathcal{B}$, such that $\tilde{f}(\pi(x)) = f(x)$ for all x in A.

(iii) the correspondence $f \to \tilde{f}$ is a topological isomorphism between Fréchet spaces $P_{\mathcal{B}}$ and A^ρ, which preserves the natural semi-norms of these spaces, and exchanges positive linear forms on A onto positive normal linear forms on $\mathcal{B}$.

(iv) for $y \in \mathcal{B}$, let $\hat{y}$ be the linear form $f \to \tilde{f}(y)$ on A^ρ. The map $y \to \hat{y}$ is a linear bijection from $\mathcal{B}$ onto $\hat{A}$" whose composite with π is the canonical injection from A into A". This map is continuous when $\mathcal{B}$ has bibounded topology and A" the strong dual topology of A^ρ. This map is bicontinuous from $\mathcal{B}$ weak onto (A", σ(A", A')).

As in $\mathbb{C}^*$-algebra theory, $\mathcal{B}$ will be called the enveloping structure of A, and π the canonical morphism from A into $\mathcal{B}$.

Proof. Let $f \geq 0$ normal on $\mathcal{B}$: the restriction of f to $\pi(A)$ is clearly of the form $f = \omega_{\xi,\xi}$, and this formula clearly holds on $\mathcal{B}$. Since any ultraweakly continuous linear form on $\mathcal{B}$ is a linear combination of positive normal linear forms, we get (i). Noting that A^+ is normal for ρ, we find (ii). Now, $\pi(A)$ satisfies lemma 7.9, and $T \to \pi(T)$ is a bijective positive map from A onto $\pi(A)$ preserving ρ-norms, thus implying (iii). Due to (i) and (iii), the weak topology of $\mathcal{B}$ is defined by the semi-norms $y \to \tilde{f}(y)$, $y \in A^\rho$, thus proving (iv).

Proposition 7.15. Let A be a $*$-algebra with condition I, $\mathcal{B}$ be the enveloping structure of A, and $\pi : A \to \mathcal{B}$ be the canonical morphism. Let ρ be a representation of A in some Hilbert space H_ρ and domain $\mathcal{D}_\rho$. There exists a unique normal representation $\tilde{\rho}$ of $\mathcal{B}$, in the Hilbert space H_ρ, admitting $\mathcal{D}_\rho$ as domain, such that $\tilde{\rho}(\pi(x)) = \rho(x)$ for all $x \in A$. The image $\tilde{\rho}(\mathcal{B})$ is the ultraweak closure of $\rho(A)$.

Proof. Two normal representations ρ_1 and ρ_2 of $\mathcal{B}$, with domain $\mathcal{D}_\rho$, such that $\rho_1(\pi(x) = \rho_2(\pi(x)) = \rho(x)$ for $x \in A$ must be ultraweakly continuous linear maps from A into $B(\mathcal{D}_\rho, \mathcal{D}_\rho)$, thus

$\rho_1 = \rho_2$ on $\mathcal{B}$ follows from ultraweak density of $\pi(A)$ into $\mathcal{B}$. The representation ρ being the direct Hilbert sum of representations associated to suitable vectors (chosen in $\mathcal{D}_\rho$), we are reduced to the case where ρ admits a cyclic vector in $\mathcal{D}_\rho$, in which case our proposition is obvious.

Thus, representations of A correspond to normal representations of the second dual A'' of A, a well-known result in $\mathbb{C}^*$-theory. It is also easily seen that a normal homomorphism π, from an ultraweakly closed space A_1 with condition II, onto a space A_2 of the same type induces an isomorphism from $(A_1)_E$ onto A_2, where E is a suitable projector of the center of A_{id}. Finally, the natural injection from A into itself extends, by proposition 7.15, to a normal representation from A'' onto A, leading to a projector E in the center of $\mathcal{B}_{id}$ (since the second dual A'' is isomorphic to $\mathcal{B}$ by [10]) such that $A = A''_E = \mathcal{B}_E$.
We will need

<u>Proposition 7.16</u>. Let A be a $\star$-algebra with condition I (acting in the Hilbert space H) and π its universal representation (acting in the Hilbert space H_π). If f is a linear form defined on A_{id} and ultraweakly continuous relatively to $H \hat{\otimes} H$, then the form $f \circ \pi^{-1}$ is ultraweakly continuous relative to $H_\pi \hat{\otimes} H_\pi$.

<u>Proof</u>. Let M be the von Neumann algebra generated by A_{id}, with strong dual M' and predual P_M, and $\mathcal{B}$ be the ultraweak closure of A with predual $P_\mathcal{B}$. By proposition 7.8, $P_\mathcal{B}$ is dense in the Banach space P_M and will here be considered as a normed space for the induced norm. Let $f_1 \geq 0$ be an extension of f to M. It follows from the bipolar theorem that the unit ball of $P_\mathcal{B}$ is dense in the unit ball of M' for weak topology $\sigma(M', M)$. Thus, there exists a net $g_\alpha \in P_\mathcal{B}$, $\alpha \in I$, directed set, such that $\|g_\alpha\| \leq \|g\|$ for all α, and $g_\alpha(T) \to f_1(T)$ for all $T \in M$ (as $\alpha \to \infty$). For $\alpha \in I$, let $(\pi_\alpha, H_\alpha, \mathcal{D}_\alpha, \zeta_\alpha)$ be the G.N.S. representation of $\mathcal{B}$ associated to g_α, with cyclic vector $\zeta_\alpha \in \mathcal{D}_\alpha$. From proposition 7.4, π_α is normal on $\mathcal{B}$, hence on M, and is completely known by its restriction to A_{id}. The representation $\pi_1 = \oplus_{\alpha \in I} \pi_\alpha$ of $\mathcal{B}$, acting in the Hilbert space $\oplus_\alpha H_\alpha$, is normal, and, for $T \in M$, $\alpha \in I$, one has

$$g_\alpha(T) = (\pi_\alpha(T)\zeta_\alpha, \alpha_\zeta) = (\pi_1(T)\zeta_\alpha, \zeta_\alpha),$$

thus $\omega_{\zeta_\alpha,\zeta_\alpha}(\pi_1(T))$ tends to $f_1(T)$, as $\alpha \to \infty$, with $T \in M$. Therefore, there exists a positive (Hilbert) ultraweakly continuous linear form φ, on the von Neumann algebra $\pi_1(M)$, such that $\varphi(\pi_1(T) = f_1(T)$, for $T \in M$. Noting that π_1 may be viewed as a subrepresentation of π, we find that $f \circ \pi^{-1}$ is (Hilbert) ultraweakly continuous on $\pi(A_{id})$.

Noting that a space with condition II and a cofinal central sequence must be a $\star$-algebra, and that conditions I or II are equivalent for a $\star$-algebra, one immediately has:

Proposition 7.17. Let A be a $\star$-algebra with condition II and cofinal central sequence A_i. The second dual of A is an algebra with condition II and a cofinal central sequence. If A is abelian, so is A''.

Using proposition 6.1°/ of [10] and p. 170 in [7], we put forward

Proposition 7.18. Let A and A_1 be spaces with condition II, with the same cofinal abelian sequence, and with $A \subset A_1$ and A ρ-dense in A_1. The strong dual spaces A^ρ and A_1^ρ are identical as Fréchet spaces. The second duals A'' and A_1'' are identical as topological vector spaces and $A'' = A_1''$ is an ultraweakly closed space with condition II and a cofinal abelian sequence.

The case $A = A_1$ is of interest, and proposition 2.2 is significant in the passage from A to A''. Using remark 1.1, we put forward

Proposition 7.19. Let A, with condition II and cofinal abelian sequence A_i, and $f = f^* \in A^\rho$, be G-invariant in sense of definition 9.1. There exists a unique couple (f_1, f_2) of positive linear forms on A, such that one has $f = f_1 - f_2$ on A, with

$$\|f\|_{A_i^2} = \|f_1\|_{A_i^2} + \|f_2\|_{A_i^2} \qquad \text{for all } i \geq 0.$$

One may deduce that the real linear space of G-invariant linear forms $f = f^*$ on A is a closed linear space of the strong dual of $A_{\mathbb{R}}$ ($A_{\mathbb{R}}$ is the real part of A).

CHAPTER 8: THE STATE SPACE

Lemma 8.1. Let $A = \cup_{i\geq 0} A_{A_i}$ be a space satisfying condition II, and f be a positive ultraweakly continuous (relative to $H \hat{\otimes} H$) linear form defined on A_{id}, such that $m_i = \sup\{f(B)\ ;\ 0 \leq B \leq A_i,\ B \text{ polynomial in } A_i^{-1}\}$ is finite for every $i \geq 0$. Then, f has a unique positive ultraweakly continuous (relatively to $\mathcal{D} \hat{\otimes} \mathcal{D}$) extension to A.

Proof. Let M be the von Neumann algebra generated by A_{id}, and $f = \Sigma_{i=1}^{+\infty} \omega_{x_i,x_i}$, $x_i \in H$, $i \geq 0$ be the unique normal extension to M of the given linear form. Up to a suitable ampliation of A, and hence of M, we are reduced to the case where $f = \omega_{x,x}$ with x in H, and it has to be shown that $x \in \mathcal{D}$. We put, for $i \geq 0$, $\mathcal{D}_i = \cap_{k\geq 0} A_i^{-k}(H)$ and $B_i = \cup_{k\geq 0} A_{id} A_i^k \times A_i^k$. Clearly, $\mathcal{D} = \cap_{i\geq 0} \mathcal{D}_i$ and f restricted to the space B_i - which satisfies condition II - remains bounded on traces on A_{id} of order-intervals of B_i, so that we may assume $A = B_i$. Take $A_i = \Delta$ for simplicity, $\bar{\Delta} = \int_1^{+\infty} \lambda\, dp_\lambda$ as the spectral representation of $\bar{\Delta}$, and let C be the $*$-algebra generated by Δ, Δ^{-1} with natural domain $\mathcal{D}$ ($= \mathcal{D}_i$). Then C consists of operators of the form $f(\Delta)$, where $f(t) = \Sigma_{n\in z} a_n t^n$ is a polynomial in variables t and t^{-1}. Elements of C_{id} are represented by functions f, such that $a_i = 0$ for $i \geq 1$, since Δ may be assumed to be unbounded. For $g(t) = \Sigma_{p\leq 0} a_p t^p$, such that $0 \leq g(t) \leq t$ for all $t \in [1, +\infty[$, one has $0 \leq g(\Delta) \leq \Delta$, thus, for a fixed value of n and of α,

$$0 \leq \int_1^n g^\alpha(\lambda)\, dp_\lambda(x, x) \leq \omega_{x,x}(g(\Delta)) \leq m_\alpha < +\infty,$$

and Lebesgue's theorem, together with the Weierstrass approximation theorem, lead to

$$\int_1^n \lambda^\alpha \, dp_\lambda(x, x) \le m_\alpha \; ; \qquad \int_1^\infty \lambda^\alpha \, dp_\lambda(x, x) \le m_\alpha \; ,$$

showing that $x \in \cap_{\alpha\geq 0}$ Dom $(\bar{\Delta}^{\alpha/2}) = \mathcal{D}$, since this holds for every $\alpha \geq 0$. The other assertions are straightforward.

Proposition 8.1. Let $A = \cup_{i\geq 0} A_{A_i}$ be a space satisfying condition II and f be a positive linear form defined on A_{id}, such that $f([0, A_i] \cap A_{id})$ is bounded, for every integer $i \geq 0$. Then, f has a unique positive linear extension to whole A.

It is sufficient to know here that the quantities m_i introduced in lemma 8.1 are finite.

Proof. We first consider the case of a $\star$-algebra A. Let π be the universal representation of A (see paragraph 7), and $\mathcal{B} = \pi(A)$ acting in the Hilbert space H_π. Noting that π preserves ρ-norms and that $f \circ \pi^{-1}(T) = \omega_{x,x}(T)$ for $T \in \mathcal{B}_{id}$ and x suitably chosen in H_π, it suffices to apply lemma 8.1 and unicity follows from the density of A_{id} in theorem 2.2. Now, let A be ultraweakly closed, with $A_1 \leq A_2 \leq \cdots \leq A_j \leq \cdots$, and for $j \geq 0$, $A_j = \cup_{k\geq 0} A_{id} A_j^k \times A_j^k$ with natural domain $\mathcal{D}_j = \cap_{k\geq 0} A_j^{-k}(H)$. By lemma 7.8, $A_j \cap L(\mathcal{D}_j)$ is a $\star$-algebra ρ_j-dense in A_j, and f, being first extended to $A_j \cap L(\mathcal{D}_j)$, then has a continuous extension f_j to A_j; positive elements in A_j being ρ_j-limits of sequences in A_{id}^+, and f_j must be positive on A_j. We need now to gather the sequence f_j, $j \geq 0$; this is direct since, for $m \leq n$, $f_n \geq 0$ restricted to A_m is a positive linear form on A_m extending f_{id}, and hence coincides with f_m, thus proving this case. Finally, for an A with condition II, we take $\mathcal{B}$ to be the σ-weak closure of A and g to be a positive extension of f to $\mathcal{B}_{id}$; observing that the supremum of lemma 8.1 is identical for f and g, then we are reduced to the situation just treated, hence the proposition.

Corollary 8.1. Let A be a space (respectively an ultraweakly closed space) satisfying condition II, f be a positive (respectively positive normal) linear form on A, and g_0 be a linear form defined on A_{id}, such that $0 \leq g_o \leq f$. Then g_o has a unique positive (respectively positive normal) linear extension g to A : one has $0 \leq g \leq f$.

Corollary 8.2. Take A with condition II, and let f be a positive linear form on A; then, f is pure iff its restriction to A_{id} is pure.

It is classical that pure linear forms on A (respectively A_{id}) correspond to extremal rays in the positive cone $(A')^+$ (respectively $A_{id})'^+$).

Proof. Let f be pure on A_{id}, and $g \geq 0$ on A such that $0 \leq g \leq f$ on A^+. Clearly, $g_{id} \leq f_{id}$ implies that $g_{id} = \lambda f_{id}$ for some $\lambda > 0$ and, due to the density of A_{id}, $g = \lambda f$, thus showing that f is pure. Conversely, let g_0 be a linear form on A_{id}, such that $0 \leq g_0 \leq f_{id}$. From proposition 8.1, g_0 has a positive linear extension to A and, obviously, $0 \leq g \leq f$ leads to $g = \lambda f$ for some $\lambda > 0$, implying that f is pure.

Proposition 8.2. Let $A = \cup_{i\geq 0} A_{A_i}$ with condition II, and let f be a linear form defined on A_{id}, such that $m_i = \sup f([-A_i, A_i] \cap A_{id})$ is bounded for every $i \geq 0$. Then f has a unique continuous linear extension f to A.

Proof. Let us assume that $A_1 \leq A_2 \leq \cdots \leq A_j \leq \cdots$ and define, for $j \geq 0$, $A_j = \cup_{k\geq 0} A_{id} A_j^k \times A_j^k$ with natural domain $\mathcal{D}_j = \cap_{k\geq 0} A_j^{-k}(H)$, which clearly satisfies condition II. We begin to treat the case where $A = A_i$ for some $i \geq 0$, and simply write $A_i = \Delta$ and, for $k \geq 0$, $m_k = \sup f([-\Delta^k, \Delta^k] \cap A_{id}) < +\infty$. Let $\hat{A}_{id}$ be the C^*-algebra completion of A_{id} and f_1 be a continuous linear extension of f to $\hat{A}_{id}$ (due to $[-Id, Id] \subset [-\Delta^k, \Delta^k]$). Since $\beta \in A_{\Delta 2k} \mapsto \Delta^{-k} \beta \Delta^{-k} \in A_{id}$ is an isometry for natural norms $(k \geq 0)$, the completion $\hat{A}_{\Delta 2k}$ coïncides with $\hat{A}_{id} \Delta^k \times \Delta^k$. For every $C = C^* \in \hat{A}_{id}$, there exists $C_0^* = C_0 \in A_{id}$ with $-Id \leq C_0 - C \leq Id$, and relation

$-\Delta^k \le C \le \Delta^k$ for a given $k \ge 0$ implies that $-2\Delta^k \le C_0 \le 2\Delta^k$, leading to $|f(C)| \le |f(C_0)| + \|f\| \le 2m_k + \|f\|$, i.e., $n_k = f([-\Delta^k, \Delta^k] \cap \hat{A}_{id}) < +\infty$ for any $k \ge 0$. There exists, for any integer $p \ge 0$, a sequence $B_n = B_n^* \in A_{id}$, ρ-convergent to Δ^p, with $B_n\Delta^{-1} = \Delta^{-1}B_n$ for $n \ge 0$ (this fact is contained in the proof of lemma 2.4). As $\mathcal{D} = \mathcal{D}_i$ is quasi-normable (see corollary 3.1), one can find, for a given value of p, an an integer k (depending on p), and a sequence $\varepsilon(n) \to 0$ such that $|(B_n - \Delta^p)^2| \le \varepsilon(n)\Delta^k$. It follows that every $\beta = B\Delta^p \times \Delta^p$ in $\hat{A}_{\Delta^{2p}}$ is the limit in the normed space $\hat{A}_{\Delta^h}$ with $h = 4p + 2)$ of the sequence $\beta_n = B\ B_n \times B_n = \beta_n^* \in A_{id}$, giving an estimation of type $|\beta_n - \beta| \le \varepsilon_1(n)\ \Delta^h$ for a suitable sequence $\varepsilon_1(n)\Phi \to 0$. Thus, $|f(\beta_n - \beta_n)| \le n_k\ (\varepsilon_1(n) + \varepsilon_1(m))$ shows that we can define $\dot{f}(\beta) = \lim_n f_1(B\ B_n \times B_n)$. Clearly, linear forms $g_n : B \in \hat{A}_{id} \to g_n(B) = f(B\ B_n \times B_n)$ are continuous for the norm of A_{id}, so that $B \in \hat{A}_{id} \to \dot{f}(\beta)$ remains linear continuous by the Banach-Steinhaus theorem, leading to $\beta \in \hat{A}_{\Delta^{2\ell}} \to \dot{f}(\beta)$ continuous on the normed space $\hat{A}_{\Delta^{2\ell}}$.

It is now easily seen that the quantity $\dot{f}(\beta)$, for β in A $(=A_i)$ is independent of the sequence β_n in A_{id} approximating β, and, the restriction of $\dot{f}$ to each $(A_{\Delta^h}, \|\ \|_{\Delta^h})$ being continuous, we find that $\dot{f}$ is continuous on $\bar{A}$, with unicity of extension due to the density of A_{id}. Finally, for a general A, we find, for every $i \ge 0$, a unique ρ_i-continuous linear form f_i on A_i extending f. When $i \le j$, the canonical injection from A_i into A_j is continuous for ρ-topologies so that f_j, restricted to A_i coïncide with f_i. Therefore, the sequence of f_i gathers into a linear extension f on A which must be ρ-continuous (hence unicity), since it is bounded on any interval $[-A_i, A_i]_A = [-A_i, A_i]_{A_i}$.

<u>Proposition 8.3</u>. Let $A = \cup_i A_{A_i}$ with condition II, and let $f \ge 0$ be a linear form on A. For every Borel function g with $g(x) \ge 1$, for $x \ge 1$, g smaller than some polynomial, such that $g(A_i) \in A$, one has $f(g(A_i)^{-1}) > 0$.

Taking $g(x) = x$, we find that $f(A_i^{-1}) > 0$.

Proof. Introducing a positive linear extension of f to the σ-weak closure, we are reduced to the case where A is σ-weakly closed. Let i be fixed. The σ_i-weak closure $\mathcal{B}$ of the $*$-algebra generated by A_i, A_i^{-1} is an abelian $*$-algebra with natural domain $\mathcal{D}_i = \cap_{k\geq 0} A_i^{-k}(H)$, and one has $C = g(A_i) \in \mathcal{B}$, with $g(A_i)\mathcal{D}_i = \mathcal{D}_i$ (the σ_i referring to $\mathcal{D}_i \hat{\otimes} \mathcal{D}_i$). Let $(\pi_f, H_f, \mathcal{D}_f, \zeta_f)$ be the G.N.S. representation of $\mathcal{B}$ associated to f; one has $\pi_f(C^{1/2})\mathcal{D}_f \subset \mathcal{D}_f$, $\pi_f(C^{-1/2})\mathcal{D}_f \subset \mathcal{D}_f$, thus $\pi_f(C^{1/2})\mathcal{D}_f = \mathcal{D}_f$ and $\zeta_f \neq 0$ implies that $\pi_f(C^{1/2})\zeta_f \neq 0$, so that $f(C) = (\pi_f(C^{1/2})\zeta_f, \pi_f(C^{1/2})\zeta_f) = \|\pi_f(C)^{1/2}\zeta_f\|^2 > 0$.

Proposition 8.4. Let $A = \cup_{i\geq 0} A_{A_i}$ be an ultraweakly closed subspace of $B(\mathcal{D}, \mathcal{D})$ with predual P_A. Let $\alpha > 0$, j be a fixed integer, and $K = \{f \geq 0,\ f \in P_A,$ such that $\|f\|_{A_i} \leq \alpha\}$. Then:

1°/ The set of $f = f^*$ in P_A, such that $\|f\|_{A_i} \leq \alpha$, is the closure in the Fréchet space P_A of the convex envelope of $K \cup (-K)$.

2°/ The set of $f = f^*$ (respectively of $f \geq 0$) in A^ρ, such that $\|f\|_{A_i} \leq \alpha$, is the closure of $K \cup (-K)$ (respectively of K) for the weak topology $\sigma(A^\rho, A)$.

Proof. For x in the zero neighborhood $V_i = \{x \in \mathcal{D} | (A_i x, x) \leq \alpha\}$ one has $\omega_{x,x} \in K$, therefore, in the duality $\langle A, A^\rho\rangle$, an element T belonging to the polar $K^o = \{T \in A | \langle f, T\rangle \geq -1$ with $f \in K\}$ of K must satisfy $(Tx, x) \geq -1$ for $x \in V_i$, i.e., $(Tx, x) \geq -\frac{1}{\alpha}(A_i x, x)$ for all $x \in \mathcal{D}$. Conversely, given T in A such that $T \geq -\frac{1}{\alpha} A_i$, one has $f(T) \geq -\frac{1}{\alpha} f(A_i)$ for $f \geq 0$ in A^ρ, and, taking f in K, we get $T \in K^o$. Consequently, the bipolar K^{oo} of K is the set of $g = g^*$ in A^ρ, such that $g(T) \geq -1$ for $T \geq -A_i/\alpha$, implying that g must be positive, and our estimation becomes equivalent to $g(A_i/\alpha) \leq 1$, i.e., $\|g\|_{A_i} \leq \alpha$. Replacing K^o by its absolute polar $\{T \in A|\ |\langle f, T\rangle| \leq 1$ for f in K$\}$, we find 2°/.

The strong dual $\tilde{A}$ of P_A coïncides as vector space with A and the bounded sets of $\tilde{A}$ agree with those of (A, ρ), so that the strong dual $\tilde{A}'$ of $\tilde{A}$ is topologically isomorphic to a subspace of the Fréchet space A^ρ. As P_A is quasi-barreled, it may be identified to a topological vector subspace of $\widetilde{A}'$, implying that the semi-norms $f \to \|f\|_{A_i}$ on A^ρ, restricted to P_A, define the topology of this Fréchet space. For 1°/, it suffices to replace the preceding duality $\langle A, A^\rho\rangle$ by $\langle A, P_A\rangle = \langle\widetilde{A}, P_A\rangle$ and $\rho(A^\rho, A)$ by $\sigma(P_A, A) = \sigma(P_A, \widetilde{A})$ and to observe that the closure of $K \subset P_A$ for $\sigma(P_A, \widetilde{A})$ coïncides with its closure in P_A, by the Minkowski theorem.

For a space A with condition II, we introduce

$$K = \{f \geq 0,\ f \text{ defined on } A_{id} \text{ such that } f(1) = 1\}$$

and

$$S = K \cap A^\rho = \{f \geq 0 \text{ on } A \text{ such that } f(1) = 1\}.$$

It is known that K is compact for $\sigma(A'_{id}, A_{id})$ and S will be definitively endowed with topology induced by that of K. The transposed map ${}^t j$ from A^ρ into the Banach space A'_{id} is injective, by theorem 2.2, so that S will be identified to a subset of K.

For T in the real part of A_{id} (respectively of A), the formula $\hat{T}(f) = f(T)$, f in K (respectively in S) defines a function $\hat{T} = \varphi(T)$ on K (respectively on S) and φ is isometric (respectively a ρ-norm preserving map, due to [11], for example) from A_{id} (respectively A) onto its image. The function $\hat{T}$ thus obtained is continuous (respectively a Borel function) on the compact set K (respectively on the Borel dense set S) when T is in A_{id} (respectively in A). For $*$-algebras with a Schwartz domain, Sherman has introduced [22] similar types of functions spaces, taking for K the unit ball of the Hilbert space.

Lemma 8.2. 1°/ S is dense in K (for topology $\sigma(A'_{id}), A_{id})$)

2°/ for f in $\varphi(A)$, there exists g in $\varphi(A)$ such that $f^2 \leq g$ on S; in particular, $\varphi(A_i)^2 \leq \varphi(A_i^2)$

3°/ for every integer $i \geq 0$, one has $\varphi(A_i)\varphi(A_i^{-1}) \geq 1$, and $\varphi(A_i^{-1}) > 0$ on S.

Proof. For $A \in A \cap L(\mathcal{D})$ and $h \in S$, it is easily seen that

$$[\varphi(A)h]^2 = h(A)^2 \leq h(A^*A)h(1) = \varphi(B)h,$$

since $B = A^*A \in \mathcal{A} \cap \mathcal{L}(\mathcal{D})$ and $h(1) = 1$. Thus, for $\beta = CA_i \times A_i$, with $C = C^*$ in $\mathcal{A}_{id}$, assuming that $-1 \leq C \leq 1$, we find that

$$[\varphi(\beta)h]^2 = h(CA_i \times A_i)^2 \leq h(A_i^2)^2 \leq \varphi(A_i^4)h,$$

thus proving 2°/. Now, let $\Omega = \{x \in \mathcal{D};\ \|x\| = 1\}$ and $A = \{\omega_{x,x}$ with x in $\Omega\}$. In the duality $\langle \mathcal{A}_{id}, \mathcal{A}_{id}^{\prime}\rangle$, the polar of A is the interval $[0, Id]$ of $\mathcal{A}_{id}$, so that A is dense for $\sigma(\mathcal{A}_{id}^{\prime}, \mathcal{A}_{id})$ in the set A^{oo} of positive linear forms g (on $\mathcal{A}$), satisfying $g(1) \leq 1$, leading to 1°/. Let $h \in S$ and $k \geq 0$ be a linear extension of h to $\mathcal{L}(\mathcal{D})$. The formula

$$1 = k(A_i^{-1/2} A_i^{1/2}) \leq [\varphi(A_i)h][\varphi(A_i^{-1})h].$$

is 3°/.

Proposition 8.5. Let $\mathcal{C}$ be the family of equicontinuous weakly closed sets (for $\sigma(\mathcal{A}^{\rho}, \mathcal{A})$) in $(\mathcal{A}^{\rho})^+$, contained in S.

1°/ any element of $\mathcal{C}$ is compact for $\sigma(\mathcal{A}_{id}^{\prime}, \mathcal{A}_{id})$

2°/ any function $f \in \varphi(\mathcal{A})$ restricted to $B \in \mathcal{C}$ is continuous. The ρ-topology of $\varphi(\mathcal{A})$ (or of $\mathcal{A}$) is the topology of uniform convergence on elements of $\mathcal{C}$.

3°/ S is a Borel set in K : elements of $\varphi(\mathcal{A})$ are Borel functions on K (chosen to be equal to $+\infty$ on K-S).

For every integer i, it is convenient to introduce $\mathcal{A}_i = \cup_{k\geq 0} \mathcal{A}_{A_i^k}$ with natural domain $\mathcal{D}_i = \cap_{k\geq 0} A_i^{-k}(H)$, and spectrum $S_i = \mathcal{A}_i^{\rho} \cap K$, which clearly satisfies condition II. For T in $\mathcal{A}_i$, we define $\varphi_i(T)$ as the function defined on K by $\varphi_i(T)(f) = T(f)$ for $f \in S_i$ and $\varphi_i(T)f = +\infty$ for $f \in K-S_i$. Such a choice is related to

Lemma 8.3. Let i be a fixed integer. For f in S_i and $n \geq 0$, one has

$$[\varphi_i(A_i^n)](f) = \sup\{\varphi(B)f \,|\, 0 \leq B \leq A_i^n\ ,\ B \in \mathcal{A}_{id}\}.$$

For g in $K-S_i$, there exists $k \geq 0$, such that

$$\sup\ \{\varphi(B)g\ ;\ 0 \leq B \leq A_i^k\ ,\ B \in \mathcal{A}_{id}\} = +\infty$$

Proof of the lemma. For f in S_i, one has $[\varphi_i(A_i^n)](f) = f(A_i^n) = \sup f([0, A_i^n] \cap \mathcal{A}_{id})$, since there exists a net (in fact, a sequence) B_α of elements in A_{id} ρ-convergent to A_i^n, and $f \geq 0$ on A_i is ρ_i-continuous. When $f \notin S_i$, there exists, from proposition 8.1, an integer k, such that

$$f([1, A_i^k] \cap A_{id}) = +\infty$$

thus proving the lemma.

Proof of the proposition. For the proof of 1°/, let $M \in C$. From Alaoglu-Birkhoff, M is compact for the weak topology $\sigma(A^\rho, A)$, and, therefore, also for the weak topology $\sigma(A^\rho, A_{id})$, by Ascoli's theorem and theorem 2.2. Since M may be considered as a subset of the dual A'_{id} of the normed space (A_{id}, $\| \|_{id}$), it follows that M is compact for $\sigma(A'_{id}, A_{id})$.

For the proof of 2°/, let C be the topology on A, or on $\varphi(A)$, of uniform convergence on the sets $B \in C$, clearly less fine than ρ. Recalling that A^+ is normal for ρ, we find that ρ-topology is the topology of uniform convergence on equicontinuous positive sets M; the positive cone $(A^\rho)^+$ is the polar of A^+, so that such M may be chosen as equicontinuous, weakly closed, and positive. Let $\mathfrak{S}$ be the family of such M, and for T in A, M in $\mathfrak{S}$, let $\langle T, M\rangle = \sup \{|f(T)| ; f \in M\}$. For a set $P \in \mathfrak{S}$, for which $0 \notin P$, the set $P' = \{f/f(1); f \in P\}$ is contained in S and equicontinuous on A. Indeed, from the continuity of the function $p \in P \to p(1)$ on the (weak) compact set P, we find $0 < m \leq M < +\infty$, so that $m \leq f(1) \leq M$ for $f \in P$, and P' is bounded on every bounded subset of (A, ρ), since, for $i \geq 0$,

$$\langle P', [-A_i, A_i]\rangle = \sup_{p' \in P'} p'(A_i)$$
$$= \sup_{p \in P} \frac{p(A_i)}{p(1)} \leq \frac{1}{m} \sup p(A_i) < +\infty,$$

which is equicontinuity.

Let T_α be a net in (A, C) with limit $T \in A$, and $0 \neq N \in \mathfrak{S}$. In order to show that $\langle T_\alpha - T, N\rangle$ tends to zero, we choose $\theta \neq 0$ in N, and put $P = \theta^\alpha + N \in \mathfrak{S}$. One clearly has, for any α,

$$|\langle T_\alpha - T, N\rangle| \leq |\langle T_\alpha - T, \theta\rangle| + |\langle T_\alpha - T, P\rangle|$$
$$\leq \theta(1)\,|\langle T_\alpha - T, \theta/\theta(1)\rangle| + \sup_{p \in P} |\, p(1)\langle T_\alpha - T, p/p(1)\rangle|.$$

The first term tends to zero, as $\alpha \to \infty$, due to $\theta/\theta(1) \in C$. The second term is smaller than

$$M \sup_{p \in P'} |\langle T_\alpha - T, p/p(1)\rangle|,$$

and, noting that P' may be replaced by its weak closure P" for $\sigma(A^\rho, A)$, we find that this term also tends to zero, since $T_\alpha \to T$ for C-topology. It follows that $C = \rho$.

For every $\beta = \beta^* = BA_i^k \times A_i^k$ in A_i (with suitable $k \geq 0$), there exists a sequence $\beta_n = \beta_n^*$ in A_{id} which is ρ-convergent to A_i. As $\mathcal{D}_i$ is quasi-normable for its natural Fréchet topology, there exists a sequence $\varepsilon(n) \to 0$ and $\ell \geq 0$ such that $|\beta_n - \beta| \leq \varepsilon(n)\Delta^\ell$ on $\mathcal{D}_i \times \mathcal{D}_i$, hence, for $f \geq 0$ on A,

$$|f(\beta_n - \beta)| = |\varphi(\beta_n - \beta)f| \leq \varepsilon(n)f(\Delta^\ell),$$

so that the functions $\varphi(\beta_n)$ tend to $\varphi(\beta)$ uniformly on positive equicontinuous sets on A_i, implying that the restriction of $\varphi(\beta)$ to such sets is a continuous function. As equicontinuous ≥ 0 sets on A induce, by restriction to A_i, similar sets, we get the second assertion.

It now follows from the proof of lemma 8.1 that there exists a sequence g_m of real polynomials, such that $0 \leq g_m(A_i^{-1}) \leq A_i$ for $m \geq 0$, with $g_m(A_i^{-1}) = B_m \in A_{id}$ ρ-converging to A_i, satisfying the following property: a positive linear form k defined on A_{id} extends continuously to A iff all the numbers $n_\ell = \sup_m k(B_m^\ell)$ are finite for all $\ell \geq 0$. Indeed, through the map π of proposition 8.1, we are reduced to the verification of this property when f is (Hilbert) ultraweakly continuous on A_{id}, and this point is contained in the estimation introduced in the preceding proof, noting that, for any $\ell \geq 0$, B_n^ℓ is ρ-converging to A_i^ℓ. Thus, for f in S_i, $\varphi(B_n^\ell)(f)$ tends to $\varphi(A_i^\ell)(f)$ and, when $f \in K - S_i$, there exists some value of ℓ such that $\sup_n \varphi(B_n)(f) = +\infty$. Noting that, for $\ell \geq 0$, $\sup_n \varphi(B_n^\ell) = \varphi(A_i^\ell)$ must be a lower semi-continuous function on K, we find that the set $K_\ell = \{f \in K |\ \varphi(A_i^\ell)f = +\infty\}$ is Borel, as well as that $\cup_\ell K_\ell = K - S_i$. Finally, $S = \cap_{i\geq 0} S_i$ shows that S is Borel in K. As seen, every $f^* = f \in \varphi(A)$ is the simple limit on S of a sequence of continuous functions on K, and hence is a Borel function.

Proposition 8.6. Let f_α (α directed set) be a net in S, converging to $f \in K-S$ for weak topology of K. There exists $A \in \mathcal{A}$, such that $\varphi(A)f_\alpha$ is an unbounded net of reals.

This proposition is condition (4') of [22].

Proof. The linear form f restricted to $(\mathcal{A}_{id}, \| \|_{id})$ is continuous, and not continuous when $\mathcal{A}_{id}$ has topology induced by $(\mathcal{A}, \rho)$, since $\mathcal{A}^+_{id}$ is ρ-dense in $\mathcal{A}^+$. Let M be the family of all f_α, $\alpha \in I$. We show that there exists an integer j, such that

$$|\langle M, [-A_j, A_j]\rangle| = \sup\{f(T);\ f \in M;\ -A_i \leq T \leq A_i\}$$

$$= \sup\{f(A_i);\ f \in M\}$$

is unbounded. Indeed, if such quantities are finite for all $j \geq 0$, then M, as well as its disked weakly closed hull M, is an equicontinuous family of linear forms on $(\mathcal{A}, \rho)$, and hence is compact for $\sigma(\mathcal{A}^\rho, \mathcal{A})$. By Ascoli's theorem, uniform structures on M induced by the weak topologies $\sigma(\mathcal{A}^\rho, \mathcal{A}_{id})$ and $\sigma(\mathcal{A}^\rho, \mathcal{A})$ coïncide. As $f_\alpha(T)$ tends to $f(T)$ (as $\alpha \to \infty$) for T in $\mathcal{A}_{id}$, we find that f_α is a Cauchy net for $\sigma(\mathcal{A}_{id}, \mathcal{A}^\rho)$, and hence for $\sigma(\mathcal{A}, \mathcal{A}^\rho)$, and compacity of M gives an element $k \in M$ adherent to the set of f_α, $\alpha \in I$ coïnciding with f on $\mathcal{A}_{id}$. It follows that k is a ρ-continuous linear extension of f to $\mathcal{A}$, which is impossible.

Proposition 8.7. At each continuous hermitian (respectively positive) linear form f on $\mathcal{A}$, it corresponds a real Radon (respectively positive) measure μ on K, such that:

1°/ $|\mu|(K - S) = 0$;

2°/ $\varphi(\mathcal{A}) \subset L^1_{\mathbb{R}}(S, d\mu)$;

3°/ for $T = T^*$ in $\mathcal{A}$, one has $f(T) = \int_S \hat{T}\, d\mu$.

Proof. As every continuous linear form f on $\mathcal{A}$ is a linear combination of positive linear forms, we are reduced to the case where $f \geq 0$. Since $\varphi \geq 0$ preserves ρ-norms of $\mathcal{A}_{\mathbb{R}}$, formula $f_0(\varphi(T)) = f(T)$ defines a positive linear form on $\varphi(\mathcal{A})$. Let $C(K)$ be the space of continuous real function on K (with its weak topology) and $\mathcal{B}$ be the set of functions on S of form $g\varphi(A_i)$, where g moves in $C(K)$ and i in $\mathbb{N}$, with (A_i) being a cofinal sequence in $\mathcal{A}^+$. Clearly, $\varphi(A_i)$ is a cofinal sequence in $\mathcal{B}$ therefore, by a theorem of Krein, f_0 has a positive extension f_1 to $\mathcal{B}$. Due to uniform continuity on K, functions belonging to $C(K)$ may be identified to their restriction to S, so that the restriction of f_1 to $C(K)$, being ≥ 0, leads to a Radon measure $\mu \geq 0$ on K, such that $f_1(g) = \int_K g\, d\mu$ for $g \in C(K)$. Recalling that the functions $\varphi(A_i)$ are lower semi-continuous on K (see lemma 8.3), we get from

$$\sup \{ \int_K g\, d\mu ;\ 0 \leq g \leq \varphi(A_i) \quad g \in C(K)\} \leq f_1(A_i)$$

that $\varphi(A_i)$ is an integrable function for μ. Now, considering μ as acting on the function space $L^{\infty}_{\mathbb{R}}(K, d\mu)$, we easily get that $\mu(K - S) = 0$. Since every f in $\varphi(A)$ is, in absolute value, smaller than a suitable homothetic of some A_i, we get 2°/, and 3°/ is an application of Lebesgue's theorem.

Before describing the consequences of proposition 8.7, we need to consider very special continuous linear form f (called localizable) on A of direct technical manipulation.

Definition 8.1. Let $A = \cup_i A_{A_i}$ be a space with condition II, and f be a linear form on A. We say that f is a localizable linear form iff there exists a projector $E \in A$, such that $E(H) \subset \mathcal{D}$ with the property $f(\beta) = f(\beta\ E \times E)$ for all $\beta \in A$.

If E is chosen such that, for all $i \geq 0$, $EA_i = A_i E$ on $\mathcal{D}$, f is said to be strictly localizable. Similarly, a projector $E \in A$ such that $E(H) \subset \mathcal{D}$, is referred to as a localizable projector.

Proposition 8.8. Let A be an ultraweakly closed space with condition II and predual P_A, with a cofinal sequence of the form $A_n = \Delta^n$ with $\Delta \geq 1$, $\Delta\mathcal{D} = \mathcal{D}$. The set of strictly localizable continuous (respectively ultraweakly continuous) linear forms on A is a linear subset dense in the Fréchet space A^{ρ} (respectively P_A).

Proof. Let $f = f^*$ in A^{ρ} (respectively in P_A). As A^+ is normal (respectively by theorem 2 of [1]), we may assume $f \geq 0$. For the proposition there exists an increasing sequence E_α of localizable projectors of A_{id}, ρ-convergent to Id with $E_\alpha \Delta^{-1} = \Delta^{-1} E_\alpha$. The linear forms $g_\alpha : \beta \in A \to g_\alpha(\beta) = f(\beta E_\alpha \times E_\alpha)$ are strictly localizable, with $g_\alpha \in P_A$ as soon as $f \in P_A$. For a fixed $i \geq 0$ one now has, for $\beta = \beta^* = BA_i \times A_i \in A$ with $-Id \leq B \leq Id$,

$$|f(B A_i \times A_i) - f(B A_i E_\alpha \times A_i E_\alpha)|$$
$$\leq |f(B(A_i - A_i E_\alpha) \times A_i)| + |f((B A_i E_\alpha \times (A_i - A_i E_\alpha))|$$
$$\leq f((A_i - A_i E_\alpha)^2)^{1/2} f(A_i^2) + f(E_\alpha A_i^2 E_\alpha) f((A_i - A_i E_\alpha)^2),$$

hence

$$\|f - g_\alpha\|_{A_i} \leq 2f(A_i^2) f((A_i - A_i E_\alpha)^2)$$

must tend to zero as $\alpha \to \infty$, since f is ρ-continuous and $(A_i - A_i E_\alpha)^2$ is ρ-convergent to zero. Since $\sup(E_1, E_2)$ is a localizable projector when E_1, E_2 are so, it follows that the sum of linear localizable forms remains localizable, hence the proposition.

Lemma 8.4. Let $f = f^*$ be a localizable continuous (respectively ultraweakly continuous) linear form on A. There exists a unique couple (g, h) of positive localizable continuous (respectively ultraweakly continuous) linear forms on A, such that $f = g - h$ on A, with $\|f\|_{id} = \|g\|_{id} + \|h\|_{id}$.

When f is strictly localizable, we easily deduce that, for every $i \geq 0$, there exists a unique couple (g_i, h_i) of ≥ 0 localizable forms on A, with $f = g_i - h_i$ on A, and $\|f\|_{A_i^2} = \|g\|_{A_i^2} + \|h\|_{A_i^2}$ the type of continuity of g_i, h_i depending on the corresponding continuity of f. Of course, A_i^2 may be replaced by A_i for an ultraweakly closed A.

Proof. We first consider the ultraweakly continuous case. The restriction f_{id} of f to M is Hilbert ultraweakly continuous, and may be viewed as an hermitian linear form on the von Neumann algebra M_E, due to theorem 4.1. Thus, there exist projectors E_0, F_0 in M_E, such that $E_0 + F_0 \leq E$ with $E_0 F_0 = 0$ satisfying, for any $B \in M_E$,

$$f(B) = f(E_0 B E_0) + f(F_0 B F_0), \qquad (*)$$

with $f(E_0 \cdot E_0) \geq 0$, $f(F_0 \cdot F_0) \leq 0$. From $E_0 \leq E$, $F_0 \leq E$, we find that $E_0(H) \subset \mathcal{D}$, $F_0(H) \subset \mathcal{D}$. Noting that $B \leq A_i^2$ implies

$B E_0 \times E_0 \leq A_i^2 E_0 \times E_0 \in M$, we see from $f(E_0 B E_0) = f(E_0(E_0 BE_0)E_0)$ that

$$\sup\{f(E_0BE_0) \ ; \ B \in M \quad 0 \leq B \leq A_i^2\} \leq f(A_i^2 E_0 \times E_0).$$

Thus, $f(E_0 \cdot E_0) = g$ has a positive linear extension to A by proposition 8.1, as well as $f(F_0 \cdot F_0) = h$, and formula (*) is satisfied for $\beta = B$ in A. Going back to the universal representation of A, we get the lemma for continuous f.

<u>Proposition 8.9</u>. Let $A = \cup_{i\geq 0} A_{A_i}$ be an ultraweakly closed space with condition II and $f = f^*$ be a ρ-continuous linear form on A whose restriction to A_{id} is Hilbert σ-weakly continuous. Then, f is σ-weakly continuous on A, i.e., f is in the predual P_A of A.

<u>Proof</u>. First treat the case of a space A as considered in proposition 8.8. It is seen that f is the ρ-limit in the strong dual A^ρ of the sequence f_n, where $f_n \equiv f(E_n \cdot E_n)$ for $n \geq 0$. Let P_M be the predual of the von Neumann algebra $M = A_{id}$; clearly, $f_n \in P_M$ and $f_n \in A^\rho$. By proposition 7.2 in [1], it suffices to show that f_n is $\sigma(A, \mathcal{D} \hat{\otimes} \mathcal{D})$ continuous. Using lemma 8.4 applied to $f_n \in A^\rho$ (one may also proceed directly), we find $g_n \geq 0$, $h_n \geq 0$ on A, such that $f_n = g_n - h_n$ on A with $\|f_n\|_{id} = \|g_n\|_{id} + \|h_n\|_{id}$, and it is known from W*-theory that the condition $f_n \in P_M$ implies $g_n \in P_M$, $h_n \in P_M$. By proposition 7.7, $g_n \in P_A$, $h_n \in P_A$, and thus $f_n \in P_A$, as is desired. For a general space A, one may use remark 1.1.1°/ since f is $\sigma(\mathcal{D}_i \hat{\otimes} \mathcal{D}_i)$-continuous on A_i for every $i \geq 0$, as shown by the preceding lines. It easily follows that f is $\sigma(\mathcal{D} \hat{\otimes} \mathcal{D}, A)$ -continuous on A (one may work directly on incomplete tensor products, replacing A by $A \otimes \mathbb{C}_{id}$, i.e., making a suitable ampliation), thus proving the proposition.

<u>Proposition 8.10</u>. Let A be ultraweakly closed with condition II and let P_M (respectively M') be the Banach space predual (respectively dual) of the Banach space $M = A_{id}$. Let E be a projector in A, such that $E(H) \subset \mathcal{D}$, and B be a subset of P_A (respectively of A^ρ), such that $f = f(E \cdot E)$ for all $f \in B$.

Then B is bounded in A^{ρ} (respectively in P_A) iff B is bounded in M' (respectively in P_M).

Proof. We are reduced to the situation where $f = f^*$ for all $f \in B$. Let $B \subset P_A$. First, by W*-theory, there exists, for every $f \in B$, positive elements f_1, f_2 in P_M, such that $f = f_1 - f_2$ on M and $\|f\|_{id} = \|f_1\|_{id} + \|f_2\|_{id}$, with $\|f_1\|_{id} = f_1(1)$, $\|f_2\|_{id} = f_2(1)$. Due to unicity, $f_1 = f_1(E \cdot E)$, $f_2 = f_2(E \cdot E)$ and, due to $EME = E\,A\,E$ by proposition 4.2, we find that $f_1, f_2 \in A^{\rho}$. By proposition 8.9, $f_1, f_2 \in P_A$. Take $\bar{f}_1$, $\bar{f}_2$, normal extensions, we may assume that $\bar{f}_1 = \bar{f}_1(E \cdot E)$ and $\bar{f}_2 = \bar{f}_2(E \cdot E)$. Then $\bar{f} \equiv \bar{f}_1 - \bar{f}_2 \in P_A$ satisfies

$$\|\bar{f}\|_{id} \le \|\bar{f}_1\|_{id} + \|\bar{f}_2\|_{id} = \bar{f}_1(1) + \bar{f}_2(1) = \|f\|_{id},$$

leading to $\|\bar{f}\|_{id} = \|f\|_{id}$, since $\bar{f}$ extends f from M to $L(H)$. It remains to show that the semi-norms

$$g \in P_{B(\mathcal{D},\mathcal{D})} \to \|g\|_{A_i,B(\mathcal{D},\mathcal{D})} = \sup_{\substack{|T| \le A_i \\ T \in B(\mathcal{D},\mathcal{D})}} |g(T)|$$

are bounded on the set $\bar{B}$ of all $\bar{f}$, with f in B. This follows from

$$\begin{aligned}
\|\bar{f}\|_{A_i,B(\mathcal{D},\mathcal{D})} &\le \|\bar{f}_1\|_{A_i,B(\mathcal{D},\mathcal{D})} + \|\bar{f}_2\|_{A_i,B(\mathcal{D},\mathcal{D})} \\
&= f_1 A_i + f_2(A_i) \\
&\le \|f_1\|_{id} \|E\,A_i\,E\| + \|f_2\|_{id} \|E\,A_i\,E\| \\
&\le \|f\|_{id} \|E\,A_i\,E\|,
\end{aligned}$$

since $\sup\{\|f\|_{id}\ ;\ f \in M\}$ is bounded. The proof is similar for $B \subset A^{\rho}$, and the converse is obvious.

We are in a position to prove the important

Theorem 8.1. Let A be a space with condition II and $f = f^*$ be a ρ-continuous (respectively σ-weakly continuous) linear form on $A = \cup_{j \ge 0} A_{A_j}$. Let $i \ge 0$ be a fixed integer : there exists a unique couple (g, h) - depending on i - of positive (respectively positive σ-weakly continuous) linear forms on A, such that $f = g - h$ on A and $\|f\|_{A_i^2} = \|g\|_{A_i^2} + \|h\|_{A_i^2}$.

In an ultraweakly closed space A, each A_i has a square-root $A_i^{1/2} \in A$, with $A_i^{1/2}\mathcal{D} = \mathcal{D}$ - due to $A_i\mathcal{D} = \mathcal{D}$ - so that we may formulate theorem 8.1 with the simpler condition $f = g - h$ and $\|f\|_{A_i} = \|g\|_{A_i} + \|h\|_{A_i}$. The notation $\|f\|_{A_i}$ is explicit in proposition 2.4.

<u>Proof</u>. Unicity of the decomposition is mentioned in proposition 2.4 for ρ-continuous f, and follows from proposition 8.9 for σ-weakly continuous (relatively to $\mathcal{D} \hat{\otimes} \mathcal{D}$) f, since g and h must be σ-weakly continuous (relatively to $H \hat{\otimes} H$) in this case. Now $\beta \to A_i\ \beta\ A_i$ and $\beta \to A_i^{-1}\ \beta\ A_i^{-1}$, being homeomorphisms inverse w.r.t. each other in the topological space (A, ρ), we get by transposition that $f \in A^\rho \to f^i \in A^\rho$ (respectively $f_i \in A^\rho$), where $f^i(\beta) = f(A_i\ \beta\ A_i)$ (respectively $f_i(\beta) = f(A_i^{-1}\ \beta\ A_i^{-1})$) for $\beta \in A$ are Fréchet homeomorphisms. Noting that $\|f^i\|_{id} = \|f\|_{A_i^2}$ (respectively $\|f_i\|_{A_i^2} = \|f\|_{id}$) due to lemma 1.1 we are thus reduced to the case where $A_i = Id$. By proposition 8.7, and keeping its notation, we get a real Radon measure μ on K satisfying $|\mu(K - S) = 0$, $\mu(\hat{T}) = f(T)$ for all $T \in A_{\mathbb{R}}$. Let $\mu = \mu^+ - \mu^-$ be its Jordan decomposition: from $|\mu| = \mu^+ + \mu^-$, we get $\mu^+(K - S) = \mu^-(K - S) = 0$, $\varphi(A) \subset L^1_{\mathbb{R}}(K, \mu^+) \equiv L^1_{\mathbb{R}}(S, \mu^+)$ and, similarly, for $L^1(S, \mu^-)$. Let g and h be the positive linear forms on A defined by $g(T) = \int_S \hat{T}\, d\mu^+$ (respectively $h(T) = \int_S \hat{T}\, d\mu^-$) for $T \in A$. Since φ is a ρ-norm preserving map from $A_{\mathbb{R}}$ onto $\hat{A} = \varphi(A_{\mathbb{R}})$ with its ρ-topology, we see that $\|g\|_{id} = \|\mu^+\| = \mu^+(1)$, $\|h\|_{id} = \|\mu^-\| = \mu^-(1)$, thus our proposition follows from $\|\mu\| = \|\mu^+\| + \|\mu^-\|$.

Completing theorem 8.1, is

<u>Proposition 8.11</u>. Let P be the real part of the predual of $B(\mathcal{D}, \mathcal{D})$ and $L^1(H, \mathcal{D})$ be the real linear subspace of all operators $\rho = \rho^* \in L(H)$ satisfying $\rho(H) \subset \mathcal{D}$ and $A_i\ \rho\ A_i$ nuclear operator for all $i \geq 0$.

1°/ for $\rho = \rho^* \in L^1(H, \mathcal{D})$ with the spectral decomposition $\rho = \Sigma_\alpha \mu_\alpha < \cdot , y_\alpha > y_\alpha$, the formula $\varphi_\rho \equiv \Sigma_\alpha \mu_\alpha \omega_{y\alpha, y\alpha}$ defines an element $\varphi_\rho \in P$. Conversely, to each $\varphi = \varphi^* \in P$, it corresponds to a unique $\rho = \rho^* \in L^1(H, \mathcal{D})$, such that $\varphi = \varphi_\rho$. One has $\varphi \geq 0$ iff $\rho \geq 0$.

2°/ $L^1(H, \mathcal{D})$, endowed with the sequence of semi-norms $\rho \in L^1(H, \mathcal{D}) \mapsto \|A_i \rho A_i\|_1$ (here, $\| \ \|_1$ stands for the usual norm-trace), is a Fréchet space topologically isomorphic to the Fréchet space P.

3°/ For $\rho \in L^1(H, \mathcal{D})$, the spectral decomposition of the nuclear operator $A_i \rho A_i$ in positive parts (i.e., $A_i\rho A_i = (A_i\rho A_i)^+ - (A_i\rho A_i)^-$ with $(A_i\rho A_i)^+(A_i\rho A_i)^- =$ $= (A_i\rho A_i)^-(A_i\rho A_i)^+ = 0$) induces in P the decomposition of φ_ρ into positive components - relative to the norm $\| \ \|_{A_i^2}$ - as established in theorem 8.1.

The reader will note that the operators $A_j|\rho|A_j$, $A_j\rho A_j$ make sense only on $\mathcal{D}$.

<u>Proof</u>. 1°/ Let $\rho = \rho^* = \Sigma_\alpha \mu_\alpha < \cdot , y_\alpha > y_\alpha$ be an element of $L^1(H); \mathcal{D})$, where (y_α) is an orthonormal basis in the Hilbert space H. From $\rho(y_\alpha) = \mu_\alpha y_\alpha$, we get $y_\alpha \in \mathcal{D}$ when $\mu_\alpha \neq 0$, due to $\rho(H) \subset \mathcal{D}$. As the finite sums of the vector states $\omega_{y_\alpha, y_\alpha}$ are in P, we need to show, by proposition 7.2 [1], that, for every $j \geq 0$, $\|\varphi_n - \varphi\|_{A_j}$ tends to zero as $n \to \infty$, where $\varphi_n \equiv \Sigma_{\alpha=1}^n \mu_\alpha \omega_{y_\alpha, y_\alpha}$. One may assume an A_j of the form A_i^2, so that we now need to estimate the quantity

$$\|\varphi_n - \varphi\|_{A_i^2} = \| \sum_{\alpha>n} \mu_\alpha \omega_{y_\alpha, y_\alpha}\|_{A_i^2} \leq \sum_{\alpha>n} |\mu_\alpha| (A_i^2 y_\alpha, y_\alpha),$$

which will tend to zero as $n \to \infty$, as soon as the convergence of $\Sigma_\alpha |\mu_\alpha| (A_i^2 y_\alpha, y_\alpha)$ is established. By the closed graph theorem, we see that ρ is a continuous linear map from the Hilbert space H into the Fréchet space $\mathcal{D}$. Now, A_i^2 is continuous from $\mathcal{D}$ Fréchet into H Hilbert, implying that $A_i^2\rho$ is a continuous bounded operator from H into itself, leading first to $A_i^2\rho(v) = \Sigma_\alpha \mu_\alpha < v, y_\alpha > A_i^2 y_\alpha$ for $v \in H$, thus $(A_i^2\rho\, y_n, y_n) = \mu_n (A_i^2 y_n, y_n)$. Noting that $A_i^2\rho A_i^2$ is a nuclear operator and A_i^{-2} bounded, we find that $A_i^2\rho$ is nuclear and, due to

$$\mathrm{Tr}|B| = \sup_{(\zeta_n)(\zeta_n')} \sum_{n=1}^{\infty} |(B\zeta_n, \zeta_n)|$$

for $B \in L(H)$, where (ζ_n) and (ζ_n') are arbitrary orthonormal systems in H, we get, with $\zeta_n = \zeta_n' = y_n$, that

$$\sum_{n=1}^{\infty} |\mu_n| (A_i^2\, y_n, y_n) \leq \mathrm{Tr}|A_i^2\rho| < +\infty,$$

leading to $\varphi \in P$. Every T in $L(\mathcal{D})$ may be written as $T = B\, A_i = B\, A_i \times \mathrm{Id}$, with $B \in L(H)$, and thus $\varphi(T) = \Sigma_\alpha\, \mu_\alpha(B\, A_i y_\alpha, y_\alpha)$. Decomposing B in positive parts iff necessary, we are reduced to $B \geq 0$ and

$$\mathrm{Tr}(T\rho) = \mathrm{Tr}(B\,(A_i\rho)) = \mathrm{Tr}(B^{1/2}(A_i\rho)B^{1/2}) =$$

$$= \sum_\alpha \mu_\alpha(B\, A_i y_\alpha, y_\alpha)$$

follows from the formula $B^{1/2}(A_i\rho)B^{1/2} = \Sigma_\alpha\, \mu_\alpha < \cdot\, , B^{1/2}y_\alpha > B^{1/2}A_i y_\alpha$.

Now, let $\varphi = \varphi^* \in P$. We first show that there exists a unique bounded operator $\rho = \rho^*$, such that $\rho(H) \subset \mathcal{D}$ and $\varphi(T) = \mathrm{Tr}(T\rho)$ for $T \in L(H)$. Using [1] theorem 2, we are reduced to $\varphi \geq 0$. The restriction φ_{id} of φ to $L(H) = B(\mathcal{D}, \mathcal{D})_{id}$ being normal, there exists a unique positive nuclear operator ρ, such that $\varphi(T) = \mathrm{Tr}(T\rho)$ for all $T \in L(H)$. Let $\rho = \Sigma_\alpha\, \mu_\alpha < \cdot\, , y_\alpha > y_\alpha$ be its spectral decomposition, where y_α is an orthonormal system in H, and $\Sigma_\alpha\, \mu_\alpha < +\infty$. Using lemma 7.7 applied to the space $\mathcal{B} = B(\mathcal{D}, \mathcal{D}) \otimes \mathbb{C}_{\ell^2(\mathbb{N})}$, with its natural domain the family of all σ-convergent sequences, we deduce that $y_\alpha \in \mathcal{D}$ for all α, such that $\mu_\alpha \neq 0$. From $\varphi(T) = \Sigma_\alpha\, \mu_\alpha(Ty_\alpha, y_\alpha)$ for $T \in L(H)$, it follows that $\varphi(B\, A_i \times A_i) = \Sigma_\alpha\, \mu_\alpha(B\, A_i y_\alpha, A_i y_\alpha)$ by the ρ-density of $L(H)$, for example. We put, for $n \geq 0$, $\rho_n \equiv \Sigma_{i=1}^n\, \mu_i < \cdot\, , y_i > y_i$. For $v \in H$ and $j \in \mathbb{N}$, one has $\rho_n\, v \in \mathcal{D}$ and, for $m \geq n$, one has

$$\|A_j(\rho_m - \rho_n)v\| = \|\sum_{i=n}^{m} \mu_i < v, v_i > A_j v_i\|$$

$$\leq \sum_{i=n}^{m} \mu_i \|v\| \, \|A_j v_i\| \, \|v_i\| \leq \sum_{i=n}^{m} \mu_i \|A_j v_i\|^2 \|v\|,$$

due to $A_j \geq 1$. As $\varphi(A_j^2) = \sum_{i=1}^{\infty} \mu_i (A_j v_i, A_j v_i)$ is finite, we find that $\rho_n(v)$ is a Cauchy sequence in the Fréchet space $\mathcal{D}$, with limit $\rho(v)$ (take $A_0 = \mathrm{Id}$), thus $\rho(v) \in \mathcal{D}$, i.e., $\rho(H) \subset \mathcal{D}$. Noting that ρ is continuous from H Hilbert into $\mathcal{D}$ Fréchet, we get, for all $v \in \mathcal{D}$,

$$(A_i \rho A_i) v = \sum_{\alpha=1}^{\infty} \mu_\alpha < v, A_i y_\alpha > A_i y_\alpha .$$

thus

$$\mathrm{Tr}(A_i \rho A_i) = \sum_{\alpha=1}^{\infty} \mu_\alpha (A_i y_\alpha, A_i y_\alpha) = \varphi(A_i^2)$$

must be finite, hence $\rho \in \mathcal{L}^1(H, \mathcal{D})$ and, obviously, $\varphi = \varphi_\rho$.

2°/ Now, a Cauchy sequence $\rho_n = \rho_n^*$ in $\mathcal{L}^1(H, \mathcal{D})$, with its semi-norms, induces, for every $j \geq 0$, a Cauchy sequence $A_j \rho_n A_j$ in the Banach space $\mathcal{L}^1(H)$ of trace class operators in H. Thus, for all $j \geq 0$, we get an element $\rho^j = (\rho^j)^* \in \mathcal{L}^1(H)$, defined by $\rho^j = \lim_n A_j \rho_n A_j$; in particular, for $j = 0$, we get $\rho^0 \in \mathcal{L}^1(H)$, due to $A_0 = \mathrm{Id}$. The natural embedding from $\mathcal{L}^1(H)$ into the Banach algebra $\mathcal{L}(H)$ being continuous, ρ^j is also the norm limit of the sequence $(A_j \rho_n A_j)$, hence $A_j^{-1} \rho^j A_j^{-1} = \rho^0$ by the norm continuity of $B \to A_j^{-1} B A_j^{-1}$. Finally, for $v \in H$, one has $\rho(v) = A_j^{-1} \rho^j A_j^{-1}(v) \in A_j^{-1}(H) = \mathrm{Dom}\, \bar{A}_j$, leading to $\rho^0(v) \in \cap_{j\geq 0} \mathrm{Dom}\, \bar{A}_j = \mathcal{D}$, i.e., $\rho^0(H) \subset \mathcal{D}$. It follows that the formula $\rho^j = A_j \rho^0 A_j$ holds on $\mathcal{D}$, so defining an element $\rho^0 \in \mathcal{L}^1(H, \mathcal{D})$, showing that $\mathcal{L}^1(H, \mathcal{D})$ is complete. Now, using an estimation in 1°/, one has for $\rho = \Sigma_\alpha \mu_\alpha < \cdot , y_\alpha > y_\alpha$,

$$\|\varphi_\rho\|_{A_i^2} \leq \|\sum_\alpha \mu_\alpha \omega_{y_\alpha, y_\alpha}\|_{A_i^2} \leq \sum_\alpha |\mu_\alpha| (A_i^2 y_\alpha, y_\alpha) \leq \| A_i \rho A_i \|_1,$$

thus showing the continuity of $\rho \to \varphi_\rho$. Since two comparable Fréchet topologies are isomorphic, we get 2°/.

3°/ This can be derived classically from W^*-theory when $A_i = \mathrm{Id}$, and one is always reduced to this case, replacing f by $f(A_i^{-1/2} \cdot A_i^{-1/2})$.

Proposition 8.12. Let $\mathcal{A}$ be ultraweakly closed in $B(\mathcal{D}, \mathcal{D})$ and $P_{\mathcal{A}}$ be its predual endowed with its Fréchet topology. Let $\mathcal{C}$ be the family of compact disked sets of the Fréchet space $\mathcal{D}$, and $(\mathcal{A}, bc)$ be the bicompact topology on $\mathcal{A}$ (i.e., uniform convergence on sets of the form $A \times B$ with A, B in $\mathcal{C}$).

1°/ The Fréchet space $P_{\mathcal{A}}$ is topologically isomorphic to the strong dual of the topological space $(\mathcal{A}, bc)$. The bicompact topology of $\mathcal{A}$ is the topology of uniform convergence on compact sets of $P_{\mathcal{A}}$.

2°/ for every $A \in \mathcal{C}$ and Radon measure μ on the compact space A, the formula

$$v(\beta) = \int_A \beta(x, x)\, d\mu(x) \qquad \text{for } \beta \in \mathcal{A}$$

defines an element v of $P_{\mathcal{A}}$. When μ moves in a bounded set of Radon measures on A (respectively positive Radon measures on A), v moves in an equicontinuous (respectively positive equicontinuous) set of $P_{\mathcal{A}}$. Conversely, any $v \in P_{\mathcal{A}}$, any equicontinuous (respectively equicontinuous positive) set on $(\mathcal{A}, bc)$ comes from this construction.

Note that equicontinuous sets on $(\mathcal{A}, bc)$ are bounded sets of the Fréchet space $P_{\mathcal{A}}$, but the converse in general fails (except for $\mathcal{D}$ Schwartz space, nuclear space, etc.).

Proof. As $\mathcal{A}$ is ultraweakly closed, $\mathcal{A}$ is as a vector space, the dual space of the Fréchet space $\mathcal{D} \hat{\otimes} \mathcal{D}/\mathcal{A}^o$, topologically isomorphic - hence identified - to $P_{\mathcal{A}}$, with $\mathcal{A}^o$, the polar of $\mathcal{A}$, in duality $< \mathcal{D} \hat{\otimes} \mathcal{D}, B(\mathcal{D}, \mathcal{D})>$. Since compact subsets of $\mathcal{D} \hat{\otimes} \mathcal{D}/\mathcal{A}^o$ are canonical images of compact subsets of $\mathcal{D} \hat{\otimes} \mathcal{D}$, we find that $(\mathcal{A}, bc)$ is the topology of compact convergence on $P_{\mathcal{A}}$. Using Mackey's theorem, we see that the dual of $(\mathcal{A}, bc)$ is, vectorially, $\mathcal{D} \hat{\otimes} \mathcal{D}/\mathcal{A}^o$: bounded sets of $(\mathcal{A}, bc)$ are subsets of $\mathcal{A}$ which are bounded on compact sets of $\mathcal{D} \hat{\otimes} \mathcal{D}$; thus, due to the principle of equicontinuity are traces on $\mathcal{A}$ of order-intervals of $B(\mathcal{D}, \mathcal{D})$, so showing that the strong dual of $(\mathcal{A}, bc)$ is $P_{\mathcal{A}}$ with topology induced by the Fréchet space $\mathcal{A}^\rho$, hence 1°/. As $\mathcal{C}$ is stable under unions and finite sums, we find from polarization equality that formulas $\beta \in \mathcal{A} \to \sup_{x \in A} |\beta(x, x)|$, with A moving in $\mathcal{C}$, are a fundamental system of semi-norms of $(\mathcal{A}, bc)$. For 2°/, let A and μ be given. The formula mentioned obviously makes sense, since continuity of each β on $\mathcal{D} \times \mathcal{D}$ ensures continuity on the diagonal of $A \times A$ - homeomorphic to A. The set of $\beta \in \mathcal{A}$, such that $|\beta(x, x)| \leq 1$ for all $x \in A$, is a zero neighborhood of $(\mathcal{A}, bc)$ on which v is bounded, hence $v \in P_{\mathcal{A}}$. Clearly, $v \geq 0$ as soon as $\mu \geq 0$. Conversely, any equicontinuous subset B in the dual of $(\mathcal{A}, bc)$ is described by an estimation of the type

$$|\varphi(\beta)| \leq M \sup_{x \in A} |\beta(x, x)|$$

for all $\varphi \in B$, and suitable A in $\mathcal{C}$, $M < +\infty$, showing that B induces an equicontinuous subset of the Banach space $C(A)$ of complex continuous functions on A, with supremum norm, thus leading to 2°/. With the help of [8], theorem 7 no. 1, and proposition 21 no. 2 §4, we find, with our notations, due to the reflexivity of $\mathcal{D}$,

Proposition 8.13. Let $\mathcal{B}$ be the family of disked, closed bounded sets of the Fréchet space $\mathcal{D}$.

1°/ Let $\varphi = \varphi^*$ be a linear form on $\mathcal{A}$. Then $\varphi \in P_{\mathcal{A}}$ iff v belongs to a space $\mathcal{D}_A \hat{\otimes} \mathcal{D}_A$ for suitable $A \in \mathcal{C}$, i.e., $v = \Sigma_i \lambda_i \omega_{x_i, y_i}$ with $(\lambda_i) \in \ell^1(\mathbb{N})$ and (x_i), (y_i) sequences contained in A.

2°/ On $\mathcal{A}$, the bicompact topology and the topology of uniform convergence on sets of the form $A \times B$ with $A \in \mathcal{C}$ and $B \in \mathcal{B}$ admit the predual $P_{\mathcal{A}}$ as a dual space.

Proposition 8.14. 1°/ Let $\mathcal{A}$ be ultraweakly closed with condition II and cofinal central sequence A_i, and let $\varphi = \varphi^* \in P_{\mathcal{A}}$. There exists $\bar{\varphi} \in P_{B(\mathcal{D}, \mathcal{D})}$, extending φ such that $\|\bar{\varphi}\|_{A_i, B(\mathcal{D},\mathcal{D})} = \|\varphi\|_{A_i, \mathcal{A}}$ for all $i \geq 0$.

2°/ Let $\mathcal{A}$ be a $*$-algebra with condition II and cofinal central sequence A_i, and let $\varphi = \varphi^* \in \mathcal{A}^\rho$. There exists a linear continuous extension $\bar{\varphi}$ of φ to $B(\mathcal{D}, \mathcal{D})$, such that $\|\bar{\varphi}\|_{A_i, B(\mathcal{D},\mathcal{D})} = \|\varphi\|_{A_i, \mathcal{A}}$ for all $i \geq 0$.

Let us recall that a natural system of semi-norms of the strong dual $\mathcal{A}^\rho$ (respectively $B(\mathcal{D}, \mathcal{D})^\rho$) of $\mathcal{A}$ (respectively of $B(\mathcal{D}, \mathcal{D})$) is given by the sequence of semi-norms

$$\varphi \in \mathcal{A}^\rho \to \|\varphi\|_{A_i, \mathcal{A}} = \sup\{|\varphi(T)| \quad |T| \leq A_i \quad T \in \mathcal{A}\}$$

$$\text{(respectively } \varphi \in B(\mathcal{D}, \mathcal{D})^\rho \to \|\varphi\|_{A_i, B(\mathcal{D},\mathcal{D})} = \sup\{ \cdots \quad T \in B(\mathcal{D}, \mathcal{D})\}\text{)}.$$

The restriction of such maps to P_A (respectively to $P_{B(\mathcal{D}, \mathcal{D})}$) is a fundamental sequence of semi-norms of the Fréchet space P_A (respectively $P_{B(\mathcal{D}, \mathcal{D})}$). When φ moves in a bounded set of P_A (respectively of A^ρ), it is straightforward that the set of $\bar{\varphi}$ moves in a bounded set of $B(\mathcal{D}, \mathcal{D})^\rho$ (respectively of $P_{B(\mathcal{D}, \mathcal{D})}$). The proposition follows from theorem 9.1 and proposition 7.14.

<u>Proposition 8.15</u>. Let A be ultraweakly closed with condition II and cofinal sequence A_i. For a bounded hermitian set M is the predual space P_A (assumed separable), there exist positive bounded sets M_1 and M_2 contained in P_A, such that $M \subset M_1 - M_2$.

CHAPTER 9: $\mathcal{G}$-INVARIANCE AND $\mathcal{G}$-TRACES

We will assume in this paragraph that A admits an abelian cofinal sequence A_i. One may consult example 7°/ of the introduction for more explanations. Let A be ultraweakly closed, satisfying condition II, with a cofinal sequence A_i, and let M be the von Neumann algebra A_{id}. We denote by P the abelian von Neumann algebra generated by all A_i^{-1} and by P_U the group of unitary operators of P. Let $\mathcal{G}$ be the group of automorphisms of M of the form $B \in M \to U B U^{-1} \in M$, with U moving in P_U. Let $M^{\mathcal{G}} = P' \cap M$ be the von Neumann algebra of fixed points of $\mathcal{G}$. Clearly, $M^{\mathcal{G}}\mathcal{D} \subset \mathcal{D}$ and $U\mathcal{D} = \mathcal{D}$ for U in P_U. Replacing P by the von Neumann algebra (denoted always by P) generated by all A_i^{-1} and the center Z of M, we get a slightly different definition of $\mathcal{G}$-invariance (with coïncidence of both notions for a normal $f \geq 0$) and nothing is changed. Such a choice of P has to be taken in definition 9.2.

Definition 9.1. Let f be a defined linear form continuous on M (respectively on A). We say that f is $\mathcal{G}$-invariant iff $f(U^{-1} T U) = f(U)$ (respectively $f(\beta(U\,\cdot, U\,\cdot)) = f(\beta))$ for all $U \in P_U$ and all $T \in M$ (respectively $\beta \in A$).

We recall that, for β in A and U in P_U, $\beta(U\,\cdot, U\,\cdot)$ denotes the sesquilinear form $(x, y) \in \mathcal{D} \times \mathcal{D} \to \beta(Ux, Uy)$. Moreover, continuity - or ultraweak continuity - on M of some linear map refers to norm topology of M or to Hilbert ultraweak topology of M (i.e., $\sigma(M, H \hat{\otimes} H)$). A similar remark holds for A relatively to ρ-topology and to the ultraweak topology of A (i.e., $\sigma(A, \mathcal{D} \hat{\otimes} \mathcal{D})$). When f is ultraweakly continuous on M (respectively on A), f is $\mathcal{G}$-invariant iff $f(A_i^{-1} B) = f(B A_i^{-1})$ (respectively $f(\beta(A_i^{-1}\,\cdot\,, \cdot)) =$
$= f(\beta(\,\cdot\,, A_i^{-1}\,\cdot)))$ for all $i \geq 0$ and $B \in M$ (respectively $\beta \in A$).

When A_n is central in A, for $n \geq 0$, every continuous linear form on A is obviously $\mathcal{G}$-invariant. Thus, for a von Neumann algebra (i.e., $M = A$), we may choose $A_i = 1$ for all $i \geq 0$ and any f on M is G-invariant. The strong dual $A_{\mathbb{R}}^{\rho}$ (respectively A^{ρ}) of $(A_{\mathbb{R}}, \rho)$ (respectively (A, ρ)) is a Fréchet space, and a natural fundamental system of semi-norms on $A_{\mathbb{R}}^{\rho}$ is given by the sequence of maps

$$f \in A_{\mathbb{R}}^{\rho} \mapsto \|f\|_{A_i} = \sup_{-A_i \leq T \leq A_i} |f(T)|.$$

Clearly, the direct sum $A^{\rho} = A_{\mathbb{R}}^{\rho} \oplus i\, A_{\mathbb{R}}^{\rho}$ is a topological sum. We shall be interested first in a special decomposition of (hermitian) G-invariant linear forms on A, well known in $\mathbb{C}^*$-theory.

<u>Lemma 9.1</u>. Let us assume that M is G-finite. Then, the unique G-invariant normal positive projection map $T \in M \mapsto T^G \in M^G$ extends uniquely to a G-invariant normal positive projection $\beta \in A \mapsto \beta^G \in \cup_{i \geq 0} M^G A_i \times A_i$.

<u>Proof</u>. Let $B = \cup_{i \geq 0} M^G A_i \times A_i$; since $A_i^{-1} \in M^G$, B is an ultraweakly closed $*$-algebra, with natural domain $\mathcal{D}$ and cofinal central sequence A_i. For every $U \in P_U$, the automorphism $\Phi_U : B \in M \to \Phi_U(B) = UBU^{-1} \in M$ extends to a normal automorphism $\tilde{\Phi}_U$ of A, by proposition 7.1; in fact, for $\beta = B\, A_i \times A_i$ with $B \in M$, one has

$$\tilde{\Phi}_U(\beta) = \beta(U^{-1} \cdot\, , U^{-1} \cdot\,) = B\, A_i U^{-1} \times A_i U^{-1} =$$

$$= U\, B\, U^{-1}\, A_i \times A_i,$$

and $\tilde{\Phi}_U(M\, A_i \times A_i) = M\, A_i \times A_i$. For β in A, let $K_0(\beta)$ the convex envelope in A of the orbit of β under the $\tilde{\Phi}_U$ automorphisms, with $U \in P_U$, and $K(\beta)$ be its ultraweak closure (relative to $\mathcal{D} \hat{\otimes} \mathcal{D}$). When $\beta \in M$, $K(\beta)$ is a bounded subset of M, and the ultraweak closure, relative to $H \hat{\otimes} H$, coïncides with its ultraweak closure relatively to $\mathcal{D} \hat{\otimes} \mathcal{D}$ by Ascoli's theorem. Let φ_i be the positive bijective map

$$B \in M \to \varphi_i(B) = B\, A_i \times A_i \in M\, A_i \times A_i = A_{A_i^2},$$

bicontinuous for weak topologies $\sigma(M, \mathcal{D} \hat{\otimes} \mathcal{D})$ and $\sigma(A_{A_i^2}, \mathcal{D} \hat{\otimes} \mathcal{D})$. Let $\beta = C\, A_i \times A_i \in A$ with $C \in M$; one has

$\varphi_i(K_0(C)) = K_0(\beta)$, hence $\varphi_i(K(C)) = K(\beta)$, thus

$$\varphi_i(C^G) = \varphi_i(K(C) \cap M^G) = \varphi_i(K(C) \cap \varphi_i(M^G) = K(\beta) \cap B_{A_i^2}$$

shows that $K(\beta) \cap B_{A_i^2}$ is reduced to $C^G A_i \times A_i$. Let $\beta = D A_j \times A_j$ be another way of writing β (for some $j \geq 0$ and $D \in M$). Choosing $k \in \mathbb{N}$ and $M < +\infty$ (we take $M = 1$), such that $A_i \leq M A_k$ and $A_j \leq M A_k$, we find an $E \in M$, such that $\beta = E A_k \times A_k$. Then, $0 \leq A_k A_i^{-1} = A_i^{-1} A_k \leq 1$, $A_i A_k^{-1} \in P$, and $C = (A_k A_i^{-1})E(A_k A_i^{-1})$ leads to

$$C^G = ((A_k A_i^{-1})E)^G A_k A_i^{-1} = A_k A_i^{-1} E^G A_k A_i^{-1} ,$$

due to $(A_i A_k^{-1})^G = A_i A_k^{-1}$ and known properties of $T \to T^G$ on M. This shows that $E^G A_k \times A_k = C^G A_i \times A_i$, leading to $D^G A_j \times A_j = C^G A_i \times A_i$, due to the similar role of D. We put $\beta^G = C^G A_i \times A_j$. The same arguments show that, for $\beta = \Sigma_{\alpha \in J} C_\alpha A_\alpha \times A_\alpha$ with J finite and $C_\alpha \in M$ for $\alpha \in J$, one has $\beta^G = \Sigma_\alpha C^G A_\alpha \times A_\alpha$. It easily follows that $B \cap K(\beta)$ is reduced to β^G. Indeed, given two elements θ_1, θ_2 in $B \cap K(\beta)$, we choose an integer $\ell \geq 0$ such that $\beta = F A_\ell \times A_\ell$, $\theta_1 \in B_{A_\ell^2}$, $\theta_2 \in B_{A_\ell^2}$ with $F \in M$, and from the properties of the map φ_ℓ, we get $B_{A_\ell^2} \cap K(\beta) = F^G A_\ell \times A_\ell$, and $\theta_1 \in K(\beta) \cap B_{A_\ell^2}$, $\theta_2 \in K(\beta) \cap B_{A_\ell^2}$ imply that $\theta_1 = \theta_2$, hence β^G can be defined as the unique element of $B \cap K(\beta)$. The formula $\beta^G = B^G A_i \times A_i$ for $\beta = BA_i \times A_i$ and the normality of each φ_i lead to the lemma.

Theorem 9.1. Let f be an ultraweakly continuous hermitian G-invariant linear form on A, and A_i be an abelian cofinal sequence in A. Then, there exists a unique couple (f_1, f_2) of positive normal G-invariant linear forms on A, such that $f = f_1 - f_2$ on A and

$$\|f\|_{A_i} = \|f_1\|_{A_i} + \|f_2\|_{A_i}$$

for every $i \geq 0$.

For unicity of the decomposition, we refer to proposition 2.4, noting that $A_i^{1/2} \in A$. Let us recall that $\|g\|_{A_i} = g(A_i)$ for any positive linear form g on A. A direct consequence of the decomposition mentioned in theorem 9.1 is that the linear set P_A^G of hermitian, ultraweakly continuous G-invariant linear forms on A is closed in the Fréchet space P_A; moreover, given an hermitian bounded set in P_A^G, there exist positive bounded sets M_1, M_2 included in P_A^G, such that $M \subset M_1 - M_2$.

Proof. We first prove the theorem when the sequence A_i is central in A. The natural domain of the commutant A' of A is $\mathcal{D}$, so that A' is the commutant of $(A', \mathcal{D})$; thus, $M\mathcal{D} \subset \mathcal{D}$. The restriction f_{id} of f to M is an hermitian, ultraweakly continuous linear form on M (for $\sigma(M, H \hat{\otimes} H)$), hence there exist the projectors E_1, E_2 in M, with $E_1E_2 = 0$, such that $f = f(E_1 \cdot E_2) - f(E_2 \cdot E_2)$ on $M = A_{id}$, with

$$\|f\|_{id} = \|f_{id}\| = \|f(E_1 \cdot E_1)\| + \|f(E_2 \cdot E_2)\|,$$

the support of $f(E_1 \cdot E_1) = f_1$ (respectively $f(E_2 \cdot E_2) = f_2$) being E_1 (respectively E_2). From $E_j\mathcal{D} \subset \mathcal{D}$, $j = 1, 2$, it follows that $f(E_j \cdot E_j)$ is ultraweakly continuous relatively to $\mathcal{D} \hat{\otimes} \mathcal{D}$, and makes sense on A, since $A(\mathcal{D}) \subset \mathcal{D}$. Thus, the formula $f = f_1 - f_2$ on A_{id} holds on whole A, due to the density of M in A, thus showing our formula for $A_i = Id$. Now, for an arbitrary integer i, one has $f(A_i \cdot\ ) = f_1(A_i \cdot\ ) - f_2(A_i \cdot\ )$ on A; it is obvious that $f(A_i \cdot\ ) = f(A_i^{1/2} \cdot A_i^{1/2})$ is ultraweakly continuous on A, and $f_j(A_i \cdot\ )$, for $j = 1, 2$ are normal on A, due to normality of the positive map $T \in A \to A_iT = A_i^{1/2}\ T\ A_i^{1/2} \in A$. Restricting to M, the support of $f_j(A_i \cdot\ )$ is seen to be equal to E_j, $j = 1, 2$, since A_i is central, thus, from $E_1E_2 = 0$, we find that

$$\|f(A_i \cdot\ )\|_{id} = \|f_1(A_i \cdot\ )\| + \|f_2(A_i \cdot\ )\|.$$

Noting that, for T in A, $-A_i \le T \le A_i$ is equivalent to $T = C\ A_i^{1/2} \times A_i^{1/2}$, with $-Id \le C \le Id$, $C \in M$, we get

$\|f(A_i \cdot)\|_{id} = \|f\|_{A_i}$ and similar relations for f_j, $j = 1, 2$, thus showing the formula $\|f\|_{A_i} = \|f_1\|_{A_i} + \|f_2\|_{A_i}$. When A_i is not necessarily central, we introduce $\mathcal{B} = \cup_{i\geq 0} M^{G} A_i \times A_i$, which is clearly an ultraweakly closed $\star$-algebra, with cofinal sequence A_i and natural domain $\mathcal{D}$, contained in $\mathcal{A}$. Considering the restriction of f to M, there exists $f_1 \geq 0$, $f_2 \geq 0$ normal on M, such that $f = f_1 - f_2$ (on M), the support E_1 to f_1 being orthogonal to the support E_2 of f_2. For $U \in P_U$ and $\Phi_U(T) = UTU^{-1}$ for $T \in M$, we find that

$$f = f \circ \Phi_U = f_1 \circ \Phi_U - f_2 \circ \Phi_U$$

and, the support of $f_1 \circ \Phi_U$ being $\Phi_U^{-1} E_1$, we find that $f_1 \circ \Phi_U$ and $f_2 \circ \Phi_U$ are orthogonal, positive linear forms on M, and, due to unicity of the orthogonal decomposition of f, we get $f_j = f_j \circ \Phi_U$ for $j = 1, 2$. It is now clear that $E = E_1 + E_2 \in M^{G}$ (hence $E\mathcal{D} \subset \mathcal{D}$ and $EA_i = A_iE$ on $\mathcal{D}$ for $i \geq 0$), that M_E is G-finite and that $f(\beta) = f(\beta(E \cdot, E \cdot))$ for every $\beta \in \mathcal{A}$ (this last assertion comes from density of M into $\mathcal{A}$). Thus, we are reduced to the case where $E = 1$.

Let g be the restriction of f to $\mathcal{B}$; then g is ultraweakly continuous relatively to $\mathcal{D} \hat{\otimes} \mathcal{D}$, so that there exists $g_1 \geq 0$, $g_2 \geq 0$ normal on $\mathcal{B}$, such that

$$\|g\|_{A_i} = \|g_1\|_{A_i} + \|g_2\|_{A_i}$$

for all $i \geq 0$. Let $\Phi : \beta \to \beta^{G} = \Phi(\beta)$ be the normal projection of lemma 9.1; then $f_1 = g_1 \circ \Phi$ and $f_2 = g_2 \circ \Phi$ are obviously positive normal on $\mathcal{A}$ and, due to ultraweak continuity of f and G-invariance, one has that $f(\beta) = f(\beta^{G})$, for $\beta \in \mathcal{A}$. Taking $\beta \in \mathcal{A}$, we obtain

$$f(\beta) = f(\beta^{G}) = g_1(\beta^{G}) - g_2(\beta^{G})$$

$$= f_1(\beta) - f_2(\beta),$$

i.e., $f = f_1 - f_2$ on $\mathcal{A}$. Moreover, $\Phi([-A_i, A_i]_{\mathcal{A}}) = [-A_i, A_i]_{\mathcal{B}}$. Indeed, $-A_i \leq \beta \leq A_i$ and $\Phi \geq 0$ imply that $-\Phi(A_i) = -A_i \leq \Phi(\beta) \leq A_i$, hence $\Phi(\beta) \in [-A_i, A_i]_{\mathcal{B}}$: conversely, for $\theta \in [-A_i, A_i]_{\mathcal{B}}$, one has $\theta = \Phi(\theta) \in [-A_i, A_i]_{\mathcal{A}}$. Finally,

$$\|f\|_{A_i} = \sup_{\beta\in[-A_i,A_i]_A} |f(\beta)| = \sup_{\beta\in[-A_i,A_i]_A} |f\, o\Phi(\beta)|$$

$$= \sup_{\gamma\in[-A_i,A_i]_B} |g(\gamma)| = \|g\|_{A_i}$$

and

$$\|g_j\|_{A_i} = g_j(A_i) = g_j\Phi(A_i) = f_j(A_i) = \|f_j\|_{A_i},$$

since $f_j \geq 0$ on A, for $j = 1, 2$. Thus, $\|f\|_{A_i} = \|f_1\|_{A_i} + \|f_2\|_{A_i}$ for all $i \geq 0$, thus proving the theorem.

Proposition 9.1. Let A be ultraweakly closed, and $f \geq 0$ be G-invariant, normal, and faithful on A. The modular group σ_t associated to the couple (A_{id}, f_{id}) extends uniquely into a one-parameter group σ_t^f of automorphisms of A. For $\beta \in A$, the function $t \to \sigma_t^f(\beta)$ is continuous from $\mathbb{R}$ into $(A, \sigma(A\ , D \hat{\otimes} D))$

Proof. Introducing, for $i \geq 0$, $A_i = \cup_{k\geq 0} A_{id} A_i^k \times A_i^k$, with natural domain $D_i = \cap_{k\geq 0} A_i^{-k}(H)$, we are reduced to the case where $A = A_i$ for some $i \geq 0$. As the G.N.S. representation π_f of (A, f) is a normal isomorphism, we are reduced to the case where $f = \omega_{\zeta,\zeta}$ with $\zeta \in D$ cyclic and separating for A_{id}. From $f(A_i^{-1}B) = f(BA_i^{-1})$ for all $B \in M$, we find that A_i^{-1} belongs to the centralizator M_f of f, i.e., $\sigma_f(A_i^{-1}) = A_i^{-1}$. The modular operator Δ, therefore, satisfies $\Delta^{it}A_i^{-1} = A_i^{-1}\Delta^{it}$, leading to $\Delta^{it}D \subset D$ and $A_i\Delta^{it} = \Delta^{it}A_i$. The unique extension of the isomorphism $\sigma_t : B \in M \to \Delta^{-it}\, B\, \Delta^{it} \in M$ is $\sigma_t^f : \beta \in A \to \beta(\Delta^{it}\,\cdot\, , \Delta^{it}\,\cdot\,) \in A$, due to the ρ-density of A_{id}. From the formula $\sigma_t^f(BA_i \times A_i) = \sigma_f^t(B)A_i \times A_i$ and lemma 5.4 [1], we get the continuity mentioned.

Proposition 9.2. Take A of proposition 9.1, and let $M = A_{id}$. The following conditions are equivalent:

1°/ For every $T \in M^+$, $T \neq 0$, there exists $f \geq 0$, G-invariant, defined, and normal on M, such that $f(T) \neq 0$.

2°/ For every $\beta \in A^+$, $\beta \neq 0$, there exists $f \geq 0$, G-invariant defined, and normal on A, such that $f(\beta) \neq 0$.

Since a normal finite trace on M is G-invariant, we have the existence of sufficiently many G-invariant normal states on a space A with finite von Neumann algebra $M = A_{id}$.

Proof. It is sufficient to show that 1°/ ⇒ 2°/. Let $f \in P_M$, where $f \geq 0$ and is G-invariant. For every $S \in P$, the linear form $B \in M \to f(S^*BS)$ is normal on M and satisfies, for $U \in P_U$, $B \in M$,

$$f(S^*UBU^{-1}S) = f(US^*BSU^{-1}) = f(S^*BS),$$

since S commutes with U and U^*, and hence is G-invariant on M. Since the natural domain $\mathcal{D}$ is essentially dense, there exists an increasing sequence E_n of projectors in P, with limit 1, such that $E_n(H) \subset \mathcal{D}$ for $n \geq 0$. As any $B \in M$ is the Hilbert ultraweak limit of the sequence E_nB, one has, for $f \in P_M$,

$$f(B) = \sum_{n\geq 0} f(B\,E_n) = \sum_{n\geq 0} f(E_nB).$$

Let $\beta = BA_i \times A_i \in A^+$, $\beta \neq 0$. Choosing an integer $n \geq 0$, such that $B\,E_n \neq 0$, we see that $BA_iE_n \times A_iE_n \neq 0$ and belongs to M. Since there exists $g \geq 0$, normal and G-invariant on M, satisfying $g(BA_iE_n \times A_iE_n) \neq 0$, we find that the linear form $h(\gamma) = g(\gamma E_n \times E_n)$, for $\gamma \in A$, is normal (due to the ultra-weak continuity of the map $\gamma \in A \to \gamma E_n \times E_n \in M$) and G-invariant on A (due to $UA_i = A_iU$ on $\mathcal{D}$ for U in P_U), and clearly $g(\beta) \neq 0$.

Lemma 9.2. Let A be an ultraweakly closed space with condition II and cofinal abelian sequence A_i. Any normal character χ defined on the ultraweakly closed abelian $*$-algebra $P = \cup_{i\geq 0} PA_i \times A_i$ extends in a positive G-invariant normal linear form f on A.

Proof. Applying remark 5.1, we see that there exists $v \in \mathcal{D}$, such that $Bv = \chi(B)v$ and $\chi(B) = \omega_{v,v}(B)$ for B in P. Taking $f = \omega_{v,v}$ on A, we find, for $\beta = BA_i \times A_i$ and $U \in P_U$,

$$\omega_{v,v}(\beta(U\ \cdot\ ,\ U\ \cdot\ )) = \omega_{v,v}(BA_iU \times A_iU)$$

$$= (BA_iUv,\ A_iUv) = \chi(U)\overline{\chi(U)}\,\omega_{v,v}(\beta),$$

hence the lemma, since $\chi(U)\ \overline{\chi(U)} = \chi(UU^*) = 1$.

Proposition 9.3. Let A be a space with condition II, and χ be a character of the abelian $*$-algebra generated by all A_i, A_i^{-1}. Any positive linear extension f of χ to A is G-invariant.

Proof. Let π be the universal representation of A with natural domain $\mathcal{D}_\pi$, B be the ultraweak closure of $\pi(A)$, and $\zeta \in \mathcal{D}_\pi$, such that $\omega_{\zeta,\zeta}(\pi(\beta)) = f(\beta)$ for all $\beta \in A$. Clearly, $\omega_{\zeta,\zeta}(\gamma) = \gamma(\zeta,\ \zeta)$ for $\gamma \in \ddot{B}$ defines the unique $\sigma(\mathcal{D}_\pi \hat{\otimes} \mathcal{D}_\pi)$ continuous extension to $\ddot{B}$ of $f \circ \pi^{-1}$. From proposition 7.3, we see that all $\pi(A_i)$, $\pi(A_i)^{-1} = \pi(A_i^{-1})$ generates an abelian $*$-algebra C with natural domain $\mathcal{D}_\pi$ and $\omega_{\zeta,\zeta}$ is a normal character χ on the σ-weak closure P of C. From the proof of lemma 9.2, we find that $B\zeta = \chi(B)\zeta$ for $B \in P$ and, by lemma 9.2 $\omega_{\zeta,\zeta}$ is a G-invariant normal linear form on B, therefore f is G-invariant on A.

Lemma 9.3. Let B be an abelian ultraweakly closed $*$-algebra with condition II and natural domain $\mathcal{D}$, and S be the set of characters of B. If $0 \neq \chi \in S$ is normal on B,

$$H_\chi = \{x \in H \quad | \ Tx = \chi(T)x \text{ for all } T \in B_{id}\}$$

$$= \{x \in \mathcal{D} \quad | \ Tx = \chi(T)x \text{ for all } T \in B\ \}$$

is a closed subspace $\neq 0$ of the Hilbert space H. When all characters on B are normal, H is the direct orthogonal Hilbert sum of the spaces H_χ; $\chi \in S$

Proof. Let χ be normal in S. The proof of theorem 5.3 shows that there exists $x \in \mathcal{D}$ with $\|x\| = 1$, such that $Tx = \chi(T)x$ for all $T \in B$, i.e., $H_\chi \neq 0$. Now, let $0 \neq v \in H$, satisfying $Tv = \chi(T)v$ for all $T \in B_{id}$. Assuming, without loss of generality, that $\|v\| = 1$, we get $\chi = \omega_{v,v}$ on B_{id}; therefore, by lemma 7.7, we find that $v \in \mathcal{D}$. It follows, from the formula $A_i^{-1}v = \chi(A_i^{-1})v$, that $v = \chi(A_i^{-1})A_iv$, and hence $A_iv = \chi(A_i)v$ for $i \geq 0$. Writing any $T \in B$ in the form $T = BA_i$, with $B \in \mathcal{B}_{id}$ and suitable $i \geq 0$, we get

$$Tv = B(\chi(A_i)v = \chi(B)\chi(A_i)v = \chi(T)v,$$

so showing the first part of the proposition and the closedness of H_χ. For second assertion we introduce $H_1 = \underset{\chi \in S}{\oplus} H_\chi, H_2 = H \ominus H_1$ and p_2 the corresponding projector in H. If $p_2 \neq 0$, there exists a normal character χ such that $\chi(p_2) \neq 0$, $\chi(A_i^{-1}) \neq 0$, for all $i \geq 0$, contradicting so the maximality of H_1, as seen in the proof of theorem 5.3.

Proposition 9.4. Let $A = \cup_{i\geq 0} A_{A_i}$ be ultraweakly closed with condition II, and $\mathcal{B}$ be the ultraweakly closed abelian $*$-algebra $\mathcal{B} = \cup_{i\geq 0} PA_i \times A_i$. If all characters on $\mathcal{B}$ are normal then A_{id} is a G-finite von Neumann algebra.

Proof. Let $T \in M^+$, $T \neq 0$, and $H = \oplus_\chi H_\chi$ be the decomposition of lemma 9.3. Applying lemma 9.2 and proposition 9.3, we see that, for any $x \in H$, the linear form $\omega_{x,x}$ on A is normal and G-invariant; thus, $(Tx, x) = 0$ for all $x \in \mathcal{D}$ and $\chi \in S$ leads to $T^{1/2}H_\chi = 0$, i.e., $T^{1/2} = 0$ is clearly impossible.

Proposition 9.5. Take $A = \cup_{i\geq 0} A_{A_i}$ of lemma 9.3, and let $\mathcal{D}$ be a Schwartz space for its natural topology. Then, the von Neumann algebra $M = A_{id}$ is discrete, semi-finite (with M' finite), and is G-finite.

Proof. Let S be the set of characters of the ultraweakly closed abelian $*$-algebra $P = \cup_{i\geq 0} PA_i \times A_i$ and, for $\varphi \in S$, H_φ the closed space of lemma 9.3, E_φ be the corresponding projector. From theorem 5.3 $Id = \Sigma_\varphi E_\varphi$, $E_\varphi \in P$, $P = \pi_{\varphi\in S}(E_\varphi P)$, so that $P' = \pi_{\varphi \in S} L(H_\varphi)$ is finite discrete, as well as $M' \subset P'$. The last assertion refers to proposition 9.4.

A notion of a G-trace is now introduced. Each unitary operator U of the von Neumann algebra $M = A_{id}$ induces an isomorphism $\Phi_U : T \in A_{id} \to UTU^{-1} \in A_{id}$, which extends uniquely (by theorem 7.1) to an isomorphism $\tilde{\Phi}_U$ from A onto $\mathcal{B} = \cup_{n\geq 0} A_{id}B_n \times B_n$, where $B_i = U A_i U^{-1}$ has domain $U(\mathcal{D})$. By proposition 7.1, $\tilde{\Phi}_U$ is an isomorphism from A onto itself iff $U(\mathcal{D}) = \mathcal{D}$. A positive linear form, invariant under certain privilegiate automorphisms Φ_U, will be called a G-trace. In this paragraph, P is the von Neumann algebra generated by all A_i^{-1} and the center Z of A_{id}.

The reader will note that a normal linear form $f \geq 0$ on A, with the property that $f(UBU^{-1}) = f(B)$ for all $B \in A_{id}$ and U in the set of all unitary operators of A_{id} with condition $U\mathcal{D} = \mathcal{D}$, must be a (finite) trace on A_{id}. Indeed, $f(AB) = f(BA)$ for all $B \in A_{id}$ and all A in the linear span F of these U, and by normality of f, A may be chosen in the σ-weak closure F^- of the space introduced. The formula $T = 1/2(V + V^*)$ for $T = T^* \in A_{id}$, with $\|T\| \leq 1$ and $V \equiv T + i(1 - T^2)^{1/2}$, gives the decomposition of $T \in L(\mathcal{D})$ into unitary V, V^* in $L(\mathcal{D}) \cap A_{id}$, implying that $V\mathcal{D} = \mathcal{D}$; thus $F \supset A_{id} \cap L(\mathcal{D})$ and, using lemma 3.3, $F^- = A_{id}$, so that f is a finite trace. It is, therefore, natural to introduce

<u>Definition 9.2</u>. A G-trace f on A (respectively on A_{id}) is a positive linear form f defined on A (respectively on A_{id}), invariant under all automorphisms Φ_U, with U in P_U^c.

Due to $UA_i^{-1} = A_i^{-1}U$ for $U \in P_U^c$, we get $U\mathcal{D} = \mathcal{D}$ and $\Phi_U(BA_i \times A_i) = BA_iU \times A_iU = U^{-1}BUA_i \times A_i$ for $B \in A_{id}$. Thus, a G-trace f on A (respectively on A_{id}) is characterized by $f(BUA_i \times A_i) = f(UBA_i \times A_i)$ for $U \in P^c$, $B \in A_{id}$, $i \in \mathbb{N}$ (respectively $i = 0$).

A linear form $f \geq 0$, defined on whole A, is a G-trace on A iff its restriction to A_{id} is a G-trace on A_{id}. Obviously, a G-trace is a G-invariant linear form, the converse being, in general, false. When P appear to be (abelian) maximal in A_{id} (i.e., $P^c = P$), the two notions mentioned coïncide. When all characters on B are normal, we find, using the notation of lemma 9.3, that $H = \oplus_{\varphi \in S} H_\varphi$ and maximality of P in the Banach algebra $L(H)$ (which obviously implies $P^c = P$), is equivalent to $\dim H_\varphi = 1$ for all $\varphi \in S$.

<u>Proposition 9.6</u>. Let A ultraweakly closed with condition II, $A_0 = A \cap L(\mathcal{D})$ and $\varphi_0 : A_{id} \to C$ be a finite trace. If φ_0 admits a positive linear extension $\varphi : A \to C$, then $\varphi(\beta(S \cdot , \cdot)) = \varphi(\beta(\cdot , S^* \cdot))$ for $\beta \in A$ and $S \in A_0$. In particular, $\varphi(ST) = \varphi(TS)$ for $T \in A_0$, $S \in A_0$.

If φ_0 is a finite normal trace on A_{id}, φ must be normal on A by proposition 7.7.

When the cofinal sequence A_i consists of central elements for A, A must be a $\star$-algebra with domain $\mathcal{D}$, and a G-trace f on A (respectively on A_{id}) is obviously a trace on A (respectively on A_{id}), i.e., $f(ST) = f(TS)$ for $S, T \in A$ (respectively $S, T \in A_{id}$).

Proof. Let $\beta = BA_i \times A_i$, with B in A_{id}. Introducing $A_i = \cup_{n\geq 0} A_{id}A_i^n \times A_i^n$ with natural domain $\mathcal{D}_i = \cap_{n\geq 0} \bar{A}_i^{-n}(H)$, we are reduced to the case $A = A_i$, since the restriction of φ to A_i is the unique positive linear extension of φ_0, due to theorem 2.1 and proposition 1.1 [9]. Let $\bar{A}_i = \int_1^\infty \lambda \, dp_\lambda$ be the spectral theorem for $\bar{A}_i$ and, for $n \in \mathbb{N}$, $E_n = \int_1^n dp_\lambda$. Due to $E_n(H) \subset \mathcal{D}_i$, we find by proposition 2.3 that $\beta_n = \beta E_n \times E_n \in A_{id}$ is ρ_i-convergent to β, as well as $\beta_n(S\,\cdot\,,\,\cdot\,) \in A$ to $\beta(S\,\cdot\,,\,\cdot\,)$ for $S \in A_0$, implying that $\varphi(\beta(S\,\cdot\,,\,\cdot\,)) = \lim_n \varphi(\beta_n(S\,\cdot\,,\,\cdot\,))$. Proceeding similarly with $\beta(\varphi(\,\cdot\,, S^*\,\cdot\,)$, it remains to prove that $\varphi(\gamma(S\,\cdot\,,\,\cdot\,)) = \varphi(\gamma(\,\cdot\,, S^*\,\cdot\,))$ for γ in A_{id} and $S \in A_0$. Conjugating theorem 2.1 and proposition 2.3, we are reduced to $S \in A_0 \cap A_{id}$, and $\gamma(S\,\cdot\,,\,\cdot\,) = \gamma S \in A_{id}$, $\gamma(\,\cdot\,, S^*\,\cdot\,) = S\gamma \in A_{id}$ imply the result, as φ_0 is a trace. Taking $\beta = T \in A_0$, it easily follows that $\varphi(ST) = \varphi(TS)$.

Proposition 9.7. Take A of proposition 9.6, and let $\mathcal{B} = \cup_{i\geq 0} PA_i \times A_i$, $\mathcal{B}^c = \cup_{i\geq 0} P^cA_i \times A_i$. Let f be a positive normal linear form on A, such that $f(UBU^{-1}) = f(B)$ for all $B \in A_{id}$ and all $U \in P_U$ (respectively $U \in P_U^c$). Then:

1°/ f is G-invariant (respectively a G-trace) on A.

2°/ $\mathcal{B}$ (respectively $\mathcal{B}^c$) is a $\star$-algebra with natural domain $\mathcal{D}$ and $f(AT) = f(TA)$ for all $A \in \mathcal{B}$ (respectively $A \in \mathcal{B}^c$) and all $T \in A_0 = A \cap L(\mathcal{D})$.

Proof. Let $\beta = BA_i \times A_i \in A$, with B bounded, and E_n be the sequence of projectors introduced in the preceding proof. It is known that $\lim_\alpha \beta E_\alpha \times E_\alpha = \beta$, and $\beta E_\alpha \times E_\alpha = A_iE_\alpha BA_iE_\alpha \in A_{id}$. Clearly, $f(A_iE_\alpha BUA_iE_\alpha = f(A_iE_\alpha UBA_iE_\alpha)$ for $U \in P_U$ and, since A_i is central in $\mathcal{B}$, we get

$$f(BUA_iE_\alpha \times A_iE_\alpha) = f(UBA_iE_\alpha \times A_iE_\alpha),$$

hence

$$f(BUA_i \times A_i) = f(UBA_i \times A_i),$$

which is G-invariance. Similarly, we get $f(BUA_i) = f(UBA_i)$. Any T in A_0 may be written $T = BA_j$, with $B \in A_{id} \cap L(\mathcal{D})$ and suitable $j \in \mathbb{N}$. Replacing A_i by A_j, we find, from $UA_j = A_jU$ on $\mathcal{D}$, that $f(BA_jU) = f(UBA_j)$, i.e., $f(TU) = f(UT)$ for $U \in P_U$, hence for $U \in P$. From $P(\mathcal{D}) \subset \mathcal{D}$, we find that $\mathcal{B}$ is a $\star$-algebra with the property mentioned. Any $A \in \mathcal{B}$ may be written as $A = CA_k$ for suitable $C \in P$ and $k \in \mathbb{N}$, and $A = \lim_\alpha AF_\alpha$, the F_α referring to $\bar{A}_k$. From $AF_\alpha \in P$ we find that $f(TAE_\alpha) = f(AE_\alpha T)$, hence $f(AT) = f(TA)$. The proof is similar for $U \in P^c_U$.

Proposition 9.8. For a space A, ultraweakly closed, of the form $A = \cup_{n\geq 0} A_{\Delta^n}$ with $\Delta \geq 1$, $\Delta\mathcal{D} = \mathcal{D}$, the following conditions are equivalent:

1°/ for any $\beta \neq 0$ in A, there exists a G-trace f on A, such that $f(\beta) \neq 0$.

2°/ for any $B \neq 0$ in $M = A_{id}$, there exists a G-trace φ on M, such that $\varphi(B) \neq 0$.

Proof. Let $f : A \to \mathbb{C}$ (respectively $f : A_{id} \to \mathbb{C}$) be a G-trace. We first show that the form $f(E \cdot E) : A \to \mathbb{C}$ is a G-trace on A (respectively on A_{id}) when E is a projector in P. Indeed, taking $B = EBE \in M$ and $U \in P^c$, we find that

$$BU\Delta \times \Delta = EBEU\Delta \times \Delta = BUE\Delta \times E\Delta = BU\Delta E \times \Delta E$$

and

$$UB\Delta \times \Delta = UEBE\Delta \times \Delta = EUBE\Delta \times \Delta = UBE\Delta \times E\Delta,$$

hence $f(\beta E \times E) = f(\beta EU \times EU)$ for β in A (respectively $\beta \in A_{id}$). Now, let E_n be an increasing sequence with $E_n(H) \subset \mathcal{D}$, $E_n \to 1$. If 2°/ holds, we choose n such that $\beta E_n \times E_n \neq 0$, and f as defined by $f(\beta) = \varphi(\beta E_n \times E_n)$ for $\beta \in A$ satisfies 1°/.

Proposition 9.9. Take A of proposition 9.4, and let $\varphi \geq 0$ be a G-invariant normal linear form on A (respectively on A_{id}), S be the set of characters on B and, for $\chi \in S$, E_χ be the projector on H_χ - see lemma 9.3. Then $\varphi = \Sigma_{\chi \in S}\, \varphi_\chi$ where φ_χ is the restriction of φ to $E_\chi\, A\, E_\chi$, and φ is a G-trace on A (respectively on A_{id}) iff φ_χ is a trace on the von Neumann algebra $E_\chi\, A\, E_\chi$ for all $\chi \in S$.

Proof. For finite I, $I \subset S$, let $F_I = \Sigma_{\chi \in I}\, E_\chi$. Due to lemma 9.3, $Id = \sup_I F_I$ and, as F_I tends to Id for $\sigma(A, D \hat{\otimes} D)$-topology we find, by proposition 2.3, that $\beta = BA_i \times A_i \in A$ is the ultraweak limit of $\beta F_I \times Id = \Sigma_\chi\, BA_i E_\chi \times A_i$. From the normality of φ, we get

$$\varphi(\beta) = \sum_\chi \varphi(\beta E_\chi \times Id) = \sum_\chi \varphi(BE_\chi A_i \times A_i)$$

and, from G-invariance,

$$\varphi(\beta) = \sum_\chi \varphi(E_\chi BA_i \times A_i) = \sum_\chi \varphi(E_\chi BE_\chi\, A_i \times A_i)$$

$$= \sum_\chi \varphi(BE_\chi A_i \times E_\chi A_i) = \sum_\chi \chi(A_i^2)\varphi(E_\chi BE_\chi),$$

hence $\varphi = \Sigma_{\chi \in S}\, \varphi_\chi$, due to $E_\chi A_i = A_i E_\chi = \chi(A_i)E_\chi$. When φ is a trace, using $P' = \pi_{\chi \in S}\, P' E_\chi = \pi_{\chi \in S} L(H_\chi)$, we find, for $U = \pi_{\chi \in S}\, U_\chi \in P^c$ and $B \in A_{id}$, that

$$\varphi(BUA_i \times A_i) = \sum_\chi \varphi(BUA_i E_\chi \times A_i E_\chi)$$

$$= \sum_\chi \chi(A_i^2)\varphi_\chi(BU) = \sum_\chi \chi(A_i^2)\varphi_\chi(E_\chi BE_\chi U_\chi)$$

and, similarly,

$$\varphi(UBA_i \times A_i) = \sum_\chi \chi(A_i^2)\varphi_\chi(U_\chi E_\chi BE_\chi).$$

From $E_\chi(H) = H_\chi$, by lemma 9.3, we find that $E_\chi\, A\, E_\chi = E_\chi\, A_{id}\, E_\chi \subset A_{id}$ by proposition 4.2, thus $E_\chi A E_\chi \subset \pi_\chi\, L(H_\chi)$ leads to $E_\chi\, A\, E_\chi \subset P^c$. Now, with a fixed χ taking B and U in $E_\chi\, A\, E_\chi$ in the preceding formula, we get $\chi(A_i^2)\varphi_\chi(UB) = \chi(A_i^2)\varphi_\chi(BU)$ i.e., φ_χ is a trace on $E_\chi AE_\chi$. Conversely, φ_χ being trace on $E_\chi AE_\chi$ for all $\chi \in S$ obviously implies that φ is a G-trace.

Conversely:

Corollary 9.2. For every $\chi \in S$, let φ_χ be a positive normal linear form (respectively a normal trace) on $E_\chi A E_\chi$.

1°/ If $\Sigma_\chi \varphi_\chi(1) < +\infty$, then $\varphi = \Sigma_\chi \varphi_\chi$ is a positive G-invariant normal linear form (respectively a normal G-trace) on A_{id}.

2°/ If $\Sigma_\chi \chi(A_j)\varphi_\chi(1) < +\infty$ for all $j \geq 0$, then $\varphi = \Sigma_\chi \varphi_\chi$ is a positive G-invariant normal linear form (respectively a normal G-trace) on A.

When S is countable, a faithful normal φ exists on A iff a normal faithful φ_0 exists on A_{id}.

The separability of H suffices to ensure that S is countable.

Proof. One has, for $T \in A_{id}$ (respectively $\beta = BA_i \times A_i \in A$), $\varphi(T) = \Sigma_\chi \varphi(E_\chi T E_\chi)$ (respectively $\varphi(T) = \Sigma_\chi \varphi(BA_i E_\chi \times A_i E_\chi) = \Sigma_\chi \chi(A_i^2)\varphi(E_\chi B E_\chi))$, implying that φ is ≥ 0 and normal on A_{id} (proposition 9.9 and proposition 7.7). Due to $P^c = \pi_{\chi \in S} E_\chi A E_\chi$, we deduce 1°/ and 2°/ with the help of computations in preceding proof. Let φ be normal faithful on A_{id}, and $\varphi = \Sigma_n \varphi_n$ be the corresponding decomposition, with S identified to $\mathbb{N}$. Clearly, $\Sigma_n (1/n!)\varphi_n$ is normal faithful on A, due to $(\chi(A_i)/n!)\ \varphi_n(1) \leq \varphi_n(1)$ for n sufficiently large (depending on i).

We now sketch

Proposition 9.10. Take A of proposition 9.4, and let φ, ψ be two normal G-invariant linear forms (respectively two normal G-traces) on A, such that $0 \leq \varphi \leq \psi$. There exists $Z \in P$ (respectively $Z \in P^c$) with $0 \leq Z \leq 1$, such that $\varphi(\beta) = \psi(\beta(Z\ \cdot\ ,\ \cdot\ ))$ for all $\beta \in A$.

Proof. Let φ, ψ be G-traces. Writing $\varphi = \Sigma_\chi \varphi_\chi$, $\psi = \Sigma_\chi \psi_\chi$, we get $\varphi_\chi \leq \psi_\chi$ for all χ, thus there exists Z_χ in the center of $E_\chi A E_\chi$, such that $\varphi_\chi = \psi_\chi(\ \cdot\ Z_\chi)$ on $E_\chi\ A_{id}\ E_\chi$. Let Z be the center of A. As $Z_\chi \in E_\chi\ Z\ E_\chi = E_\chi\ Z$, we easily see, from $Z_\chi\ Z_\eta = 0$ for $\chi \neq \eta$, that $Z = \Sigma_\chi Z_\chi \in P$ satisfies our requirement. The proof is similar for G-invariance.

Proposition 9.11. Let $A = \cup_i A_{A_i}$ be ultraweakly closed, as in proposition 9.7, and $f \geq 0$ be normal faithful on A with the associated G.N.S. representation $(\pi_f, H_f, \mathcal{D}_f, \zeta_f)$. Then, $\pi_f(BA_i \times A_i) = \chi(A_i^2)\, \pi_f(B)$ for all $B \in A_{id}$ iff the restriction of f to P is a character χ.

It follows that the G.N.S. representation associated to a G-invariant normal $f \geq 0$ on A (we may assume f faithful replacing A by EAE with $E \in P^c$) is described by the formula $\pi_f(BA_i \times A_i) = \oplus_\chi\, \chi(A_i^2)\pi_{f_\chi}(B)$ for $B \in A_{id}$ and $f = \Sigma_\chi\, f_\chi$.

Proof. Take $f \geq 0$ on A, with restriction to P equal to a character χ, and $i \geq 0$ fixed. From $f(A_i^2)f(A_i^{-2}) = \chi(A_i^2)\chi(A_i^{-2}) = 1$ or, equivalently, $(\pi_f(A_i)^2\zeta_f, \zeta_f)(\pi_f(A_i^{-1})^2\zeta_f, \zeta_f) = 1$, we deduce that $\pi_f(A_i^{-1})\zeta_f = \chi(A_i^{-1})\zeta_f$. By proposition 7.5, $\pi_f(A_{id}) = \pi_f(A)_{id}$ is a von Neumann algebra N and, by the nature of G.N.S. construction, $\zeta_f \in \mathcal{D}_f$ is a cyclic vector for N; since f is faithful, ζ_f is also separating. For $T \in N'$, one has $T\mathcal{D}_f \subset \mathcal{D}_f$ and $T\pi_f(A_i)^{-1}) = \pi(A_i^{-1})T$ on $\mathcal{D}_f$. Thus

$$\chi(A_i^{-1})T\zeta_f = T\pi_f(A_i^{-1})\zeta_f = \pi_f(A_i^{-1})\, T\, \zeta_f,$$

i.e., $\pi(A_i^{-1})v = \chi(A_i^{-1})v$ for $v \in N'\zeta_f$,

leading to $\pi(A_i)^{-1} = \chi(A_i^{-1})Id_{H_f}$ due to the boundedness of $\pi(A_i^{-1})$. As $\pi(A_i)$ is the inverse on $\mathcal{D}_f$ of $\pi(A_i^{-1})$, we get $\pi(A_i)v = \chi(A_i)v$, $v \in \mathcal{D}_f$, and hence, for $B \in A_{id}$, $\pi_f(BA_i \times A_i) = \pi_f(B)\pi_f(A_i) \times \pi_f(A_i) = \chi(A_i^2)\pi_f(B)$. The converse is straightforward, since $\pi_f(Id) = Id_{H_f}$ leads to $\pi_f(A_i) = \chi(A_i)Id_{H_f}$.

Corollary 9.3. Let L be the convex set of positive G-traces f on A, such that $f(1) \leq 1$. Then, extremal points $\neq 0$ of L may be identified to characters of the abelian $*$-algebra P generated by all A_i, A_i^{-1}.

CHAPTER 10: OTHER COMMUTATION THEOREMS

In order to clarify this Chapter, we will be concerned first with an ultraweakly closed space A with condition II, of the form $A = \cup_{n\geq} \; M\Delta^n \times \Delta^n$ (where $\Delta \geq \mathrm{Id}$ is a self-adjoint operator affiliated to a given von Neumann algebra M). This will enable us to deal more simply with an ultraweakly closed space A with condition II and cofinal abelian sequence. This point of view is motivated by remark 1.1.1/. The notations used in this Chapter are those of propositions 3.12 and 3.13, and summed up in

Proposition and definitions 10.1. Let M be a von Neumann algebra with commutant M', $\Delta \geq \mathrm{Id}$ (respectively $\nabla \geq \mathrm{Id}$) be a self-adjoint operator η to M (respectively to M'), and $\mathcal{D}_\Delta$, $\mathcal{D}_\nabla$, $\mathcal{D}_{\Delta\nabla}$ be as introduced in proposition 3.12. Take $A = \cup_{n\geq 0} M\Delta^n \times \Delta^n$ (respectively $B = \cup_{n\geq 0} M'\nabla^n \times \Delta^n$) with natural domain $\mathcal{D}_\Delta$ (respectively $\mathcal{D}_\nabla$), let $\Delta = \int_1^\infty \lambda \, dp_\lambda$ (respectively $\nabla = \int_1^\infty \lambda \, dq_\lambda$) be the spectral decomposition, and $E_\alpha = \int_1^\alpha dp_\lambda$ (respectively $F_\alpha = \int_1^\alpha dq_\lambda$) for $\alpha \geq 1$.

1°/ $A \cap L(\mathcal{D}_\Delta)$ coïncide with $A \cap L(\mathcal{D}_{\Delta\nabla})$. The involutive algebra $m_\Delta = \cup_{\alpha\geq 1} E_\alpha M E_\alpha$ is contained in the involutive algebra $M \cap L(\mathcal{D}_{\Delta\nabla})$, is ρ-dense and $\sigma(\mathcal{D}_\Delta \hat{\otimes} \mathcal{D}_\Delta)$ dense in A and generates the von Neumann algebra M.

2°/ A symmetric formulation of 1°/, where B (respectively F_α) replaces A (respectively E_α), Δ replaces ∇, and the converse.

The notation A^Δ (respectively B^∇) may appear better than A (respectively B) and will be used when necessary, however, it can make the comprehension of the text difficult.

Here, $L(\mathcal{D}_{\Delta\nabla})$ is the $\star$-algebra of all operators T, such that Dom T $\supset \mathcal{D}_{\Delta\nabla}$, Dom $T^+ \supset \mathcal{D}_{\Delta\nabla}$ and $T\mathcal{D}_{\Delta\nabla} \subset \mathcal{D}_{\Delta\nabla}$, $T^+\mathcal{D}_{\Delta\nabla} \subset \mathcal{D}_{\Delta\nabla}$.

Proof. We show that $A \cap L(\mathcal{D}_\Delta) \subset A \cap L(\mathcal{D}_{\Delta\nabla})$. Let $B \in A_{id} \cap L(\mathcal{D}_\Delta)$; one has $B\mathcal{D}_\Delta \subset \mathcal{D}_\Delta$, $B\mathcal{D}_\nabla \subset \mathcal{D}_\nabla$ due to $M\mathcal{D}_\nabla \subset \mathcal{D}_\nabla$ and $B\nabla^n = \nabla^n B$ on $\mathcal{D}_\nabla$ for $n \geq 0$ by theorem 1.2 applied to B.

For v in $\mathcal{D}_{\Delta\nabla}$, i.e., $v \in \mathcal{D}_\Delta \cap \mathcal{D}_\nabla$ and $\nabla^n v \in \mathcal{D}$ for all $n \geq 0$, we get $Bv \in \mathcal{D}_\Delta \cap \mathcal{D}_\nabla$ and $\nabla^n Bv = B\nabla^n v \in B(\mathcal{D}_\Delta) \subset \mathcal{D}_\Delta$ hence $Bv \in \mathcal{D}_{\Delta\nabla}$ i.e. $B \in L(\mathcal{D}_{\Delta\nabla})$. From $\Delta^k \mathcal{D}_{\Delta\nabla} = \mathcal{D}_{\Delta\nabla}$, we find that $\Delta^k \in L(\mathcal{D}_{\Delta\nabla})$ for $k \in \mathbb{Z}$. Since $A \cap L(\mathcal{D}_\Delta)$ is a $\star$-algebra with condition I, natural domain $\mathcal{D}_\Delta$ and cofinal sequence Δ^n, as shown in the proof of theorem 7.3 given $T \in A \cap L(\mathcal{D}_\Delta)$, one has $T^*T \in \mathcal{D}_\Delta$, leading to an estimation of the type $\|Tx\| \leq M\|\Delta^n x\|$ for suitable $M < +\infty$ and $n \in \mathbb{N}$, for all $x \in \mathcal{D}_\Delta$, thus $T = B\Delta^n$ with $B \in M \cap L(\mathcal{D}_\Delta)$, immediately implying $T \in L(\mathcal{D}_{\Delta\nabla})$. Conversely, let $T^* = T \in A \cap L(\mathcal{D}_{\Delta\nabla})$. Due to $T\mathcal{D}_{\Delta\nabla} \subset \mathcal{D}_{\Delta\nabla}$, T induces a continuous linear map from the Fréchet space $\mathcal{D}_{\Delta\nabla}$ into itself. Thus, for every integer $n \geq 0$, there exists $p \in \mathbb{N}$ and $M < +\infty$, such that $\|(\Delta\nabla)^n Tv\| \leq M\|(\Delta\nabla)^p v\|$ for all $v \in \mathcal{D}_{\Delta\nabla}$: we may assume $p \geq n$ since $\Delta\nabla \geq \mathrm{Id}$. The formula $(\nabla\Delta)^n = \nabla^n\Delta^n = \Delta^n\nabla^n$ holds on $\mathcal{D}_{\Delta\nabla}$, hence by replacing v by $\nabla^{-n}v$, and using $\nabla^{-1}\mathcal{D}_{\Delta\nabla} = \mathcal{D}_{\Delta\nabla}$, we find that

$$\|\Delta^n\nabla^n\, T\, \nabla^{-n}v\| \leq M\|\Delta^p\, \nabla^{-n}v\|$$

for all $v \in \mathcal{D}_{\Delta\nabla}$. As $T \in A$, $T(Bx, y) = T(x, B^*y)$ for all $x, y \in \mathcal{D}_\Delta$, $B \in M'$ by theorem 1.2, hence $T(\nabla^{-n}x, y) = T(x, \nabla^{-n}y)$ for $x, y \in \mathcal{D}_{\Delta\nabla}$ leads to $(T\nabla^{-n}x, y) = (Tx, \nabla^{-n}y) = (\nabla^{-n}Tx, y)$ since T is a linear operator on $\mathcal{D}_{\Delta\nabla}$ and $\nabla^{-1}\mathcal{D}_{\Delta\nabla} = \mathcal{D}_{\Delta\nabla}$ (noting that $A \subset B(\mathcal{D}_\Delta, \mathcal{D}_\Delta) \subset B(\mathcal{D}_{\Delta\nabla}, \mathcal{D}_{\Delta\nabla})$), i.e., $T\nabla^{-n} = \nabla^{-n}T$ on $\mathcal{D}_{\Delta\nabla}$, therefore

$$\|\Delta^n Tv\| \leq M\|\nabla^{-n}\Delta^p v\| \leq M\|\Delta^p v\|$$

for all $v \in \mathcal{D}_{\Delta\nabla}$, due to $\nabla^{-1} \leq \mathrm{Id}$, implying that T is a continuous linear map from the metric space $\mathcal{D}_{\Delta\nabla}$, endowed with the topology induced by the Fréchet space $\mathcal{D}_\Delta$, into itself.

Finally, by proposition 3.12, 3°/, T extends continuously from the Fréchet space $\mathcal{D}_\Delta$ into itself, leading to $T \in L(\mathcal{D}_\Delta)$, showing that $A \cap L(\mathcal{D}_\Delta) = A \cap L(\mathcal{D}_{\Delta\nabla})$. Since $\sup_\alpha E_\alpha = \mathrm{Id}$, m_Δ obviously generates M and is ρ-dense in A, as established in the proof of lemma 7.8, for example. Ultraweak density relative to $\mathcal{D}_\Delta \hat{\otimes} \mathcal{D}_\Delta$ follows directly. Finally, $m_\Delta \subset A_{id} \cap L(\mathcal{D}_\Delta) = A_{id} \cap L(\mathcal{D}_{\Delta\nabla})$, hence the proposition.

Definition 10.2. The commutant of $\mathcal{A} = \cup_{n\geq 0} M\Delta^n \times \Delta^n$ in $B(\mathcal{D}_\nabla, \mathcal{D}_\nabla)$ denoted $(\mathcal{A}', \mathcal{D}_\nabla)$ is the set of $\beta \in B(\mathcal{D}_\nabla, \mathcal{D}_\nabla)$, such that

$$\beta(Ax, y) = \beta(x, A^*y)$$

for all $x, y \in \mathcal{D}_{\Delta\nabla}$ and all A in an involutive subalgebra $\mathcal{C}$ of $M \cap L(\mathcal{D}_\Delta)$ which generates M as a von Neumann algebra (i.e., $\mathcal{C}'' = M$).

Due to theorem 10.1, this commutant does not depend on the choice of $\mathcal{C}$ in $M \cap L(\mathcal{D}_\Delta)$. For example, we may take $\mathcal{C} = m_\Delta$ or $\mathcal{C} = \mathcal{A}_{id} \cap L(\mathcal{D}_\Delta)$.

Theorem 10.1. 1°/ The commutant of $\mathcal{A}$ in $B(\mathcal{D}_\nabla, \mathcal{D}_\nabla)$ is $\cup_n M'\nabla^n \times \nabla^n$ ($\equiv \mathcal{B}$).

2°/ The $*$-algebra $\mathcal{A} \cap L(\mathcal{D}_{\Delta\nabla})$ (respectively $\mathcal{B} \cap L(\mathcal{D}_{\Delta\nabla})$) satisfies condition I, has cofinal sequence Δ^n (respectively ∇^n), is ρ-dense in $\mathcal{A}$ (respectively $\mathcal{B}$), and $\mathcal{A} \cap L(\mathcal{D}_{\Delta\nabla})$ commutes with $\mathcal{B} \cap L(\mathcal{D}_{\Delta\nabla})$ on $\mathcal{D}_{\Delta\nabla}$.

3°/ Given β in $B(\mathcal{D}_\Delta, \mathcal{D}_\Delta)$, one has $\beta \in \mathcal{B}$ iff $\beta(Ax, y) = \beta(x, A^*y)$ for all $A \in \mathcal{A} \cap L(\mathcal{D}_{\Delta\nabla})$, $x, y \in \mathcal{D}_{\Delta\nabla}$.

4°/ A similar formulation of 3°/, obtained by exchanging $\mathcal{A}$ with $\mathcal{B}$, Δ with ∇ ··· holds.

When $\Delta = Id$ and $\nabla \neq Id$, $\mathcal{A}$ is exactly the von Neumann algebra M, and the commutant of M in $B(\mathcal{D}_\nabla, \mathcal{D}_\nabla)$ is $\cup_{n\geq 0} M'\nabla^n \times \nabla^n$ ($\neq$ M in general). When $\Delta \neq Id$ and $\nabla = Id$, the commutant of $\mathcal{A}$ in $B(\mathcal{D}_\nabla, \mathcal{D}_\nabla) = L(H)$ is M'.

Proof. Let $A \in L(\mathcal{D}_\Delta) \cap \mathcal{A} = L(\mathcal{D}_{\Delta\nabla}) \cap \mathcal{A}$ by proposition 10.1. As $\nabla^{-n} \in M'$ we get, from theorem 1.2 applied to $\mathcal{A}$, $A\nabla^{-n} = \nabla^{-n}A$ on $\mathcal{D}_\Delta$; thus, by proposition 10.1.1°/, using $A \in L(\mathcal{D}_{\Delta\nabla})$ and $\nabla^{-1}\mathcal{D}_{\Delta\nabla} = \mathcal{D}_{\Delta\nabla}$, we get $A\nabla^{-n} = \nabla^{-n}A$ on $\mathcal{D}_{\Delta\nabla}$ and, composing each side of this equality by ∇^n left and right, we obtain $\nabla^n A = A\nabla^n$ on $\mathcal{D}_{\Delta\nabla}$. Similarly, for $B \in L(\mathcal{D}_\nabla) \cap \mathcal{B}$, we find that $\Delta^n B = B\Delta^n$ on $\mathcal{D}_{\Delta\nabla}$.

Let $\beta = C\nabla^k \times \nabla^k \in B(\mathcal{D}_\nabla, \mathcal{D}_\nabla)$ for suitable $k \geq 0$ and $B \in L(H)$ satisfying $\beta(Ax, y) = \beta(x, A^*y)$ for all $A \in \mathcal{C}$, x, y in $\mathcal{D}_{\Delta\nabla}$. Taking $A = A^*$, the formula

$$(C\nabla^k Ax,\ \nabla^k y) = (C\nabla^k x,\ \nabla^k Ay)$$

becomes

$$(CA\nabla^k x,\ \nabla^k y) = (C\nabla^k x,\ A\nabla^k y);$$

hence, due to $\nabla^k \mathcal{D}_{\Delta\nabla} = \mathcal{D}_{\Delta\nabla}$,

$$(CAx,\ y) = (Cx, (Ay)$$

for all $A \in \mathcal{C}$, and hence $C \in M'$, since $\mathcal{C}$ generates M. Thus, $\beta = C\nabla^k \times \nabla^k \in \cup_i M'\nabla^i \times \nabla^i$. Conversely, given $C \in M'$, one easily see that $\beta = C\nabla^k \times \nabla^k \in B(\mathcal{D}_\nabla, \mathcal{D}_\nabla)$ satisfies $\beta(Ax, y) = \beta(x, A^*y)$ for all $A \in \mathcal{C}$, $x, y \in \mathcal{D}_{\Delta\nabla}$.

By proof of theorem 7.3 and proposition 10.1, $\mathcal{A} \cap L(\mathcal{D}_\Delta) = \mathcal{A} \cap L(\mathcal{D}_{\Delta\nabla})$ must be a $*$-algebra with condition I, ρ-dense in $\mathcal{A}$ by proposition 2.1; and similarly for $\mathcal{B} \cap L(\mathcal{D}_\nabla)$. Let $A \in L(\mathcal{D}_\Delta) \cap \mathcal{A}$ and $B \in L(\mathcal{D}_\nabla) \cap \mathcal{B}$. Choosing n sufficiently large, such that $A^*A \leq M\Delta^{2n}$ with $M < +\infty$, we find that $A_0 \in M \cap L(\mathcal{D}_{\Delta\nabla})$, such that $A = A_0\Delta^n$; similarly, $B = B_0\nabla^k$ with $B_0 \in M' \cap L(\mathcal{D}_{\Delta\nabla})$. Since the operators introduced are in $L(\mathcal{D}_{\Delta\nabla})$, one has $AB = A_0\Delta^n B_0\nabla^k = A_0B_0\Delta^n\nabla^k$ and $BA = B_0\Delta^n A_0\nabla^k =$ $= B_0A_0\nabla^k\Delta^n$; thus $AB = BA$, due to $A_0B_0 = B_0A_0$ and $\Delta^n\nabla^k = \nabla^k\Delta^n$, hence 2°/.

Now, let $\beta = C\nabla^k \times \nabla^k \in \mathcal{B}$ (with $C \in M'$) and $A \in \mathcal{A} \cap L(\mathcal{D}_{\Delta\nabla})$. Then $A = A_0\Delta^n$ with $A_0 \in M \cap L(\mathcal{D}_{\Delta\nabla})$. One has, for $x, y \in \mathcal{D}_{\Delta\nabla}$, since $A_0(\mathcal{D}_{\Delta\nabla}) \subset \mathcal{D}_{\Delta\nabla}$,

$$\begin{aligned}\beta(Ax,\ y) &= (C\nabla^k A_0\Delta^n x,\ \nabla^k y) = (CA_0\nabla^k\Delta^n x,\ \nabla^k y)\\ &= (CA_0\Delta^n\nabla^k x,\ \nabla^k y) = (A_0C\Delta^n\nabla^k x,\ \nabla^k y)\\ &= (A_0\Delta^n C\nabla^k x,\ \nabla^k y),\end{aligned}$$

due to $\Delta^n C = C\Delta^n$ on $\mathcal{D}_\Delta$ by theorem 1.2. In the same way,

$$\begin{aligned}\beta(x,\ A^*y) &= (C\nabla^k x,\ \nabla^k\Delta^n A_0^* y) = (C\nabla^k x,\ \Delta^n\nabla^k A_0^* y)\\ &= (C\nabla^k x,\ \Delta^n A_0^*\nabla^k y)\end{aligned}$$

and, noting that $C\nabla^k(\mathcal{D}_{\Delta\nabla}) \subset C(\mathcal{D}_{\Delta\nabla}) \subset C(\mathcal{D}_\Delta) \subset \mathcal{D}_\Delta$, we find that

$$\beta(x, A^*y) = (\Delta^n C\nabla^k x, A_0^*\nabla^k y) = (A_0\Delta^n C\nabla^k x, \nabla^k y),$$

hence $\beta(Ax, y) = \beta(x, A^*y)$, as is wished. The converse is immediately obvious, since $C \subset L(\mathcal{D}_\Delta) \cap A$.

Definition 10.3. The commutant of A in $B(\mathcal{D}_{\Delta\nabla}, \mathcal{D}_{\Delta\nabla})$, denoted by $(A', \mathcal{D}_{\Delta\nabla})$, is the set of $\beta \in B(\mathcal{D}_{\Delta\nabla}, \mathcal{D}_{\Delta\nabla})$, such that

$$\beta(Ax, y) = \beta(x, A^*y)$$

for all $x, y \in \mathcal{D}_{\Delta\nabla}$ and all A in an ultraweakly dense, involutive subalgebra C of A, with $C \subset A \cap L(\mathcal{D}_\Delta)$.

When $\nabla = \mathrm{Id}$, our definition coïncides, with that of [1] by proposition 1.2. To be more complete, the commutant of A in $B(\mathcal{D}_\Delta, \mathcal{D}_\Delta)$ has been shown in [1] to be a $*$-algebra $\bigcup_{n\geq 0} M'(\Delta_Z)^n \times (\Delta_Z)^n$ with a cofinal central sequence $(\Delta_Z)^n = (\Delta^n)_Z$ (due to lemma A.3 in [1]) and natural domain $D = \bigcap_{n\geq 0}\Delta_Z^{-n}(H)$ ([1], theorem 3), where Δ_Z is the largest lower bound among positive self-adjoint operators $\leq \Delta$ on $\mathrm{Dom}\,\Delta \times \mathrm{Dom}\,\Delta$, and affiliated to the center Z of M.

Lemma 10.1. Δ_Z satisfies $\Delta_Z(\mathcal{D}_{\Delta\nabla}) = \mathcal{D}_{\Delta\nabla}$ and $\Delta_Z^{-1}(\mathcal{D}_{\Delta\nabla}) = \mathcal{D}_{\Delta\nabla}$.

Proof. From $M\mathcal{D}_\nabla \subset \mathcal{D}_\nabla$, $M'\mathcal{D}_\Delta \subset \mathcal{D}_\Delta$, we find that $(M \cap M')(\mathcal{D}_\Delta \cap \mathcal{D}_\nabla) \subset \mathcal{D}_\Delta \cap \mathcal{D}_\nabla$. As $\Delta_Z^{-1}\nabla^{-n} = \Delta^{-n}\nabla_Z^{-1}$ as bounded operators, one has $\Delta_Z^{-1}(\mathcal{D}_\nabla) \subset \mathcal{D}_\nabla$ and $\Delta_Z^{-1}\nabla^k = \nabla^k\Delta_Z^{-1}$ on $\mathcal{D}_\nabla$ for every $k \geq 0$. For $x \in \mathcal{D}_{\Delta\nabla}$, one has $\Delta_Z^{-1}x \in \mathcal{D}_\Delta \cap \mathcal{D}_\nabla$, since $\Delta_Z^{-1} \in M \cap M'$ and $\nabla^n\Delta_Z^{-1}x = \Delta_Z^{-1}\nabla^n x \in \Delta_Z^{-1}(\mathcal{D}_\Delta) \subset \mathcal{D}_\Delta$, hence $\Delta_Z^{-1}(\mathcal{D}_{\Delta\nabla}) \subset \mathcal{D}_{\Delta\nabla}$. Knowing that $\Delta_Z \leq \Delta$ on $\mathcal{D} \times \mathcal{D}$, and that Δ_Z commutes with Δ on $\mathcal{D}_\Delta$ (in fact $\Delta_Z\mathcal{D}_\Delta = \mathcal{D}_\Delta$ $\Delta(\mathcal{D}_\Delta) = \mathcal{D}_\Delta$), we find $\Delta_Z^\varepsilon\Delta^\eta = \Delta^\eta\Delta_Z^\varepsilon$ on $\mathcal{D}_\Delta$ for $\varepsilon = \pm 1$, $\eta = \pm 1$, implying that $\|\bar{\Delta}_Z x\| \leq \|\bar{\Delta}x\|$ for $x \in \mathrm{Dom}\,\bar{\Delta} \subset \mathrm{Dom}(\bar{\Delta}_Z)$, and showing that $\bar{\Delta}_Z\bar{\Delta}^{-1}$ makes sense on $\bar{\Delta}(\mathrm{Dom}\,\bar{\Delta}) = H$ and is thus a bounded operator in the Hilbert space H, with norm ≤ 1. Moreover, as $\Delta_Z^{-1} \in M \cap M'$, we get $\bar{\Delta}^{-k}(\bar{\Delta}_Z\bar{\Delta}^{-1}) = (\bar{\Delta}_Z\bar{\Delta}^{-1})\bar{\Delta}^{-k}$ for every

$k \geq 0$, and one can easily check that $\bar{\Delta}_Z\bar{\Delta}^{-1} \in M$, since Δ is affiliated to M. Noting that, for $n \geq 1$,
$\Delta^{-n}\nabla^{-n}(H) \subset \Delta^{-n}(H) \subset \Delta^{-1}(H) = \text{Dom}\ \bar{\Delta}$, we see that
$\bar{\Delta}_Z(\Delta^{-n}\nabla^{-n}(H))$ is a well-defined quantity, and

$$\bar{\Delta}_Z(\Delta^{-n}\nabla^{-n}(H)) \subset \Delta^{-n+1}\nabla^{-n}\bar{\Delta}_Z\bar{\Delta}^{-1}(H) \subset \Delta^{-n+1}\nabla^{-n+1}(H)$$

for $n \geq 1$. From $\mathcal{D}_{\Delta\nabla} \subset \mathcal{D}_{\Delta}$, $\Delta_Z(\mathcal{D}_{\Delta\nabla})$ makes sense, and

$$\Delta_Z(\mathcal{D}_{\Delta\nabla}) \subset \bigcap_{n\geq 1} \bar{\Delta}_Z(\Delta^{-n}\nabla^{-n}(H)) \subset \bigcap_{n\geq 1} \Delta^{-n+1}\nabla^{-n+1}(H) = \mathcal{D}_{\Delta\nabla},$$

i.e., $\Delta_Z(\mathcal{D}_{\Delta\nabla}) \subset \mathcal{D}_{\Delta\nabla}$. From the injectivity of Δ_Z, Δ_Z^{-1}, we get

$$\Delta_Z(\mathcal{D}_{\Delta\nabla}) = \mathcal{D}_{\Delta\nabla}, \qquad \Delta_Z^{-1}(\mathcal{D}_{\Delta\nabla}) = \mathcal{D}_{\Delta\nabla}.$$

Theorem 10.2. 1°/ For $n \in \mathbb{N}$, $(\Delta_Z\nabla)^n \geq \text{Id}$ is an essentially self-adjoint operator on $\mathcal{D}_{\Delta\nabla}$ $(\Delta_Z\nabla)^n\mathcal{D}_{\Delta\nabla} = \mathcal{D}_{\Delta\nabla}$ and
$D_\nabla \equiv \cap_{n\geq 0}(\Delta_Z\nabla)^{-n}(H) \supset \mathcal{D}_{\Delta\nabla}$.

2°/ The commutant of A in $B(\mathcal{D}_{\Delta\nabla}, \mathcal{D}_{\Delta\nabla})$ is $\cup_{n\geq 0}M'(\Delta_Z\nabla)^n \times (\Delta_Z\nabla)^n \equiv (A', \mathcal{D}_{\Delta\nabla})$, and is an ultraweakly closed space relative to $\mathcal{D}_{\Delta\nabla} \hat{\otimes} \mathcal{D}_{\Delta\nabla}$ and to $D_\nabla \hat{\otimes} D_\nabla$, satisfying condition II.

3°/ $M' \cap L(\mathcal{D}_{\Delta\nabla})$ (respectively $M' \cap L(D_\nabla)$) is ρ-dense in $(A', \mathcal{D}_{\Delta\nabla})$.

Since D_∇ is the Fréchet space completion of $\mathcal{D}_{\Delta\nabla}$ endowed with the sequence of semi-norms $x \in \mathcal{D}_{\Delta\nabla} \to \|(\Delta_Z\nabla)^n x\|$, we obviously have $(A', \mathcal{D}_{\Delta\nabla}) = (A', D_\nabla)$. We note in 3°/ that $(A', D_\nabla) \cap L(D_\nabla)$ is a $\star$-algebra with condition I and cofinal sequence $(\Delta_Z\nabla)^n$, with natural domain D_∇, as seen through proposition 10.1 applied to $\cup_{n\geq 0}M'(\Delta_Z\nabla)^n \times (\Delta_Z\nabla)^n$. Due to $(\Delta_Z\nabla)(\mathcal{D}_{\Delta\nabla}) = \mathcal{D}_{\Delta\nabla}$, established in 1°/, we also get a similar formulation for $(A', \mathcal{D}_{\Delta\nabla}) \cap L(\mathcal{D}_{\Delta\nabla})$, except that $\mathcal{D}_{\Delta\nabla}$ is not the natural domain of this $\star$-algebra. The $\star$-algebras $A \cap L(\mathcal{D}_{\Delta\nabla})$ and $(A', \mathcal{D}_{\Delta\nabla}) \cap L(\mathcal{D}_{\Delta\nabla})$ are commuting each other, as easily checked.

The commutant $A' = (A', \mathcal{D}_{\Delta})$ i.e., A' as calculated in

$B(\mathcal{D}_\Delta, \mathcal{D}_\Delta)$, has introduced the domain $D = \cap_{n\geq 0}\Delta_Z^{-n}(H)$ and corresponds, in theorem 10.2, to $\nabla \in \mathbb{C}$ Id. Thus, our notation D_∇ seems consistent with the preceding notation. Finally, due to $\Delta_Z \geq \mathrm{Id}$, the commutant $(A', \mathcal{D}_\Delta)$ is contained in $(A', \mathcal{D}_{\Delta\nabla})$.

<u>Proof</u>. From $\Delta_Z(\mathcal{D}_{\Delta\nabla}) = \mathcal{D}_{\Delta\nabla}$, $\Delta(\mathcal{D}_{\Delta\nabla}) = \mathcal{D}_{\Delta\nabla}$, and $\Delta_Z^{-1}\Delta^{-1} = \Delta^{-1}\Delta_Z^{-1}$, we find that $\Delta_Z\Delta = \Delta\Delta_Z$ on $\mathcal{D}_{\Delta\nabla}$, thus $(\Delta_Z\Delta)^n = \Delta_Z^n\Delta^n$ and $1 \leq \Delta_Z$, $\Delta_Z^{1/2}(\mathcal{D}_{\Delta\nabla}) = \mathcal{D}_{\Delta\nabla}$ implies that $\Delta_Z^n\Delta^n \geq \mathrm{Id}$ and this operator is essentially self-adjoint on $\mathcal{D}_{\Delta\nabla}$, since its range is dense in H. By lemma 1.2 $(A', \mathcal{D}_{\Delta\nabla})$ is closed for $\sigma(\mathcal{D}_{\Delta\nabla} \hat{\otimes} \mathcal{D}_{\Delta\nabla})$-topology.

Let $\beta \in B(\mathcal{D}_{\Delta\nabla}, \mathcal{D}_{\Delta\nabla})$: we choose C in $L(H)$ and $k \in \mathbb{N}$, such that $\beta = C(\Delta\nabla)^k \times (\Delta\nabla)^k$. By lemma 1.2, we may assume $\beta = \beta^*$, i.e., $C = C^*$. For $\beta \in (A', \mathcal{D})$, the formula $\beta(Ax, y) = \beta(x, A^*y)$ with $A \in \mathcal{C}$ and $x, y \in \mathcal{D}_{\Delta\nabla}$, becomes

$$(C\Delta^k\nabla^k Ax, \Delta^k\nabla^k y) = (C\Delta^k\nabla^k x, \Delta^k\nabla^k A^* y),$$

hence

$$(C\Delta^k A\nabla^k x, \Delta^k\nabla^k y) = (C\Delta^k\nabla^k x, \Delta^k A^*\nabla^k y),$$

due to $A\nabla^k = \nabla^k A$ on $\mathcal{D}_\Delta$ by theorem 1.2, and hence on $\mathcal{D}_{\Delta\nabla}$, since $\mathcal{C} \subset L(\mathcal{D}_{\Delta\nabla})$ and $\nabla^k(\mathcal{D}_{\Delta\nabla}) = \mathcal{D}_{\Delta\nabla}$. Replacing x by $\nabla^k x$ and y by $\nabla^k y$, we find, for all $x, y \in \mathcal{D}_{\Delta\nabla}$, that

$$(C\Delta^k Ax, \Delta^k y) = (C\Delta^k x, \Delta^k A^* y).$$

By proposition 4 [10], $\gamma = C\Delta^k \times \Delta^k \in B(\mathcal{D}_\Delta, \mathcal{D}_\Delta)$ and $\gamma(Ax, y) = \gamma(x, A^*Y)$ for all $x, y \in \mathcal{D}_{\Delta\nabla}$, and $A \in \mathcal{C}$ shows that γ must be in $L(\mathcal{D}_\Delta)$ and that γ commutes to $A \cap L(\mathcal{D}_\Delta)$ by proposition 1.2. Thus, $\gamma = B\Delta_Z^n \times \Delta_Z^n$ for suitable n and $B \in M'$. It follows that $\beta = \gamma(\nabla^k \cdot, \nabla^k \cdot) = B\Delta_Z^n\nabla^k \times \Delta_Z^n\nabla^k$ on $\mathcal{D}_{\Delta\nabla}$. Thus, by lemma 10.1, $\beta = \Delta^{-2k}\nabla^{-n}B\nabla^{-n}(\Delta_Z\nabla)^{n+k}\times(\Delta_Z\nabla)^{n+k}$ on $\mathcal{D}_{\Delta\nabla}$, due to $\Delta_Z\nabla = \nabla\Delta_Z$ on $\mathcal{D}_{\Delta\nabla}$ and, since $\Delta_Z^{-2k}\nabla^{-n}B\nabla^{-n} \in M'$ (as $B \in M'$, $\nabla^{-1} \in M'$, $\Delta_Z^{-1} \in M \cap M'$), we find that $\beta \in \cup_{n\geq 0}M'(\Delta_Z\Delta)^n\times(\Delta_Z\Delta)^n$. Conversely, for C in M', $k \in \mathbb{N}$, we can easily check that $\beta = C(\Delta_Z\nabla)^k \times (\Delta_Z\nabla)^k$ satisfies the formula $\beta(Ax, y) = \beta(x, A^*y)$

for $x, y \in \mathcal{D}_{\Delta\nabla}$ and $A \in \mathcal{C}$, as can be seen by reading the preceding lines the other way round. It follows that $(A', \mathcal{D}_{\Delta\nabla})$ satisfies condition II, and hence must be $\sigma(D_\nabla \hat{\otimes} D_\nabla)$-closed.

Let $\beta = C(\nabla\Delta_Z)^k \times (\nabla\Delta_Z)^k$ with $C \in M'$ and $k \geq 0$. Clearly, $\beta \in L(\mathcal{D}_{\Delta\nabla})$ iff $C \in L(\mathcal{D}_{\Delta\nabla})$, due to $\nabla(\mathcal{D}_{\Delta\nabla}) = \mathcal{D}_{\Delta\nabla}$ and $\Delta_Z(\mathcal{D}_{\Delta\nabla}) = \mathcal{D}_{\Delta\nabla}$. Take F_α, $\alpha \geq 1$, as introduced in proposition 10.1. By lemma 2.4, we can get an estimate of the type $\|(\nabla - \nabla F_\alpha)x\| \leq 1/\alpha \|\nabla^2 x\|$ for all $x \in \mathcal{D}_\nabla$ implying that, for any $k \geq 0$,

$$\|(\nabla^k - \nabla^k F_\alpha x\| \leq 1/\alpha \|\nabla^{k+1} x\|, \; x \in \nabla^k(\mathcal{D}_\nabla) = \mathcal{D}_\nabla$$

thus we are led to

$$|(C\nabla^k x \times \nabla^k x) - (C\nabla^k F_\alpha x \times \nabla^k F_\alpha x)| \leq \varepsilon(\alpha)(\nabla^{2k+4} x, x)$$

for $x \in \mathcal{D}_\nabla$, where $\varepsilon(\alpha)$ is a sequence tending to zero as $\alpha \to \infty$. Taking x in $\mathcal{D}_{\Delta\nabla}$, and replacing x by $\Delta_Z^k x$, by lemma 10.1 we get

$$|(C(\nabla\Delta_Z)^k x, (\nabla\Delta_Z)^k x) - (C\nabla^k F_\alpha \Delta_Z^k x, \nabla^k F_\alpha \Delta_Z^k x)|$$
$$\leq \varepsilon(\alpha)(\nabla^{k+2}\Delta_Z^k x, \nabla^{k+2}\Delta_Z^k x).$$

By the closed graph theorem, $\nabla^k F_\alpha = F_\alpha \nabla^k$ is a bounded operator in the Hilbert space H, as well as

$$C_1 \equiv C\nabla^k F_\alpha \times \nabla^k F_\alpha = \nabla^k F_\alpha C \nabla^k F_\alpha \in L(\mathcal{D}_{\Delta\nabla}) \cap \mathcal{B},$$

by proposition 10.1.1°/. Finally, $C_1(\mathcal{D}_{\Delta\nabla}) \subset \mathcal{D}_{\Delta\nabla}$ implies that $C_1\Delta_Z^k \times \Delta_Z^k = \Delta_Z^k C_1 \Delta_Z^k \in L(\mathcal{D}_{\Delta\nabla})$ by lemma 10.1, and this element obviously lies in $\mathcal{B}$. Noting that Δ_Z^{-1} and its spectral projection G_α belongs to $M \cap M'$, we find that the spectral projectors of $\bar{\Delta}_Z$ (or of Δ_Z^{-1}) are in $L(\mathcal{D}_{\Delta\nabla})$, thus $C_2 \equiv \Delta_Z^k C_1 \Delta_Z^k$ is the ρ-limit of the sequence $C_2 G_\alpha \times G_\alpha \in L(\mathcal{D}_{\Delta\nabla})$, i.e.,

$$|C_2 - C_2 G_\alpha \times G_\alpha| \leq \varepsilon_1(\alpha)\Delta_Z^{2k+4}$$

on $\mathcal{D}_{\Delta\nabla} \times \mathcal{D}_{\Delta\nabla}$, where $\varepsilon_1(\alpha)$ is a suitable sequence tending to zero as $\alpha \to \infty$. Finally, on $\mathcal{D}_{\Delta\nabla} \times \mathcal{D}_{\Delta\nabla}$,

$$|\beta - C_2 G_\alpha \times G_\alpha| \leq |\beta - C_1 \Delta_Z^k \times \Delta_Z^k| + |C_2 - C_2 G_\alpha \times G_\alpha|$$
$$\leq \varepsilon(\alpha) \nabla^{k+2} \Delta_Z^k \times \nabla^{k+2} \Delta_Z^k + \varepsilon_1(\alpha) \Delta_Z^{k+2} \times \Delta_Z^{k+2},$$

due to $\nabla\Delta_Z = \Delta_Z\nabla$ on $\mathcal{D}_{\Delta\nabla}$, $\Delta_Z \geq \mathrm{Id}$ and $\mathrm{Id} \times \mathrm{Id} \leq \nabla^{k+2} \times \nabla^{k+2}$, thus showing that β is the limit of the sequence $C_2 G_\alpha \times G_\alpha$, $\alpha \in \mathbb{N}$, hence the density of $M' \cap L(\mathcal{D}_{\Delta\nabla})$ as established, since one easily see that $C_2 G_\alpha \times G_\alpha$ is in M'. The density of $M' \cap L(D_\nabla)$ follows from theorem 2.1 as applied to $\nabla\Delta_Z$ in place of ∇.

Corollary 10.1. Let M be a factor. The commutants of $\cup_n M\Delta^n \times \Delta^n$ calculated in $B(\mathcal{D}_\nabla, \mathcal{D}_\nabla)$ and in $B(\mathcal{D}_{\Delta\nabla}, \mathcal{D}_{\Delta\nabla})$ coïncide.

This follows from $\Delta_Z \in \mathbb{C}\,\mathrm{Id}$.

Corollary 10.2. In theorem 10.2, let us assume that $(\nabla x, x) \leq (\Delta x, x)$ for all $x \in \mathcal{D}_{\Delta\nabla}$. Then, the commutant of $\cup_n M\Delta^n \times \Delta^n$ in $B(\mathcal{D}_{\Delta\nabla}, \mathcal{D}_{\Delta\nabla})$ coïncides with the commutant in $B(\mathcal{D}_\Delta, \mathcal{D}_\Delta)$ (and thus, by [1], is a $*$-algebra with natural domain D).

From $\Delta_Z(\mathcal{D}_{\Delta\nabla}) = \mathcal{D}_{\Delta\nabla}$, $\Delta_Z\Delta = \Delta\Delta_Z$ on $\mathcal{D}_{\Delta\nabla}$, we find that $(\Delta_Z\nabla)^n \leq \Delta_Z^{2n} \leq \Delta_Z^{2n}\nabla^{2n} \leq (\Delta_Z\nabla)^{2n}$ on $\mathcal{D}_{\Delta\nabla}$ thus $B(\mathcal{D}_{\Delta\nabla}, \mathcal{D}_{\Delta\nabla}) = B(D, D) \equiv B(\mathcal{D}_\Delta, \mathcal{D}_\Delta)$, where $D = \cap_{n\geq 1}\Delta_Z^{-n}(H)$; the corollary follows from proposition 3.12 and lemma 1.2.1°/.

Remarks. 1°/ The commutant of A in $B(\mathcal{D}_{\Delta\nabla}, \mathcal{D}_{\Delta\nabla})$ may be viewed, by theorem 10.2, as the commutant of A in $B(\mathcal{D}_{\nabla_1}, \mathcal{D}_{\nabla_1})$, as studied in theorem 10.1, where ∇_1 of this theorem has to be replaced by $\nabla_1 = \Delta_Z\nabla$.

2°/ When ∇^{-1} is chosen in the center of M, the commutant $\cup_n M'\nabla^n \times \nabla^n$ of A calculated in $B(\mathcal{D}_\Delta, \mathcal{D}_\Delta)$ is an ultraweakly closed $*$-algebra with condition II and natural domain $\mathcal{D}_\nabla$.

3°/ For an A with cofinal central sequence (i.e., $\Delta^{-1} \in M \cap M'$), one obviously has $\Delta_Z = \Delta$.

Now, using proposition 3.13 and its notation, we can succinctly find

Theorem 10.3. 1°/ The commutant of $A = \cup_{n\geq 0} M\Delta_n \times \Delta_n$ calculated in $B(\mathcal{D}_{(\nabla)}, \mathcal{D}_{(\Delta)})$ is $\cup_{n\geq 0} M'\nabla_n \times \nabla_n \equiv \mathcal{B}$. The $*$-algebra $A \cap L(\mathcal{D}_{(\Delta\nabla)})$ (respectively $\mathcal{B} \cap L(\mathcal{D}_{(\Delta\nabla)})\cdots$) satisfies condition II, has a cofinal sequence Δ_n, and is ultraweakly dense in A; these two $*$-algebras commute each other on $\mathcal{D}_{(\Delta\nabla)}$.

2°/ The commutant of A calculated in $B(\mathcal{D}_{(\Delta\nabla)}, \mathcal{D}_{(\Delta\nabla)})$ is $\cup_{n\geq 0} M'(\Delta_n)_Z\nabla_n \times (\Delta_n)_Z\nabla_n$ and $M' \cap L(\mathcal{D}_{(\Delta\nabla)})$ is ultraweakly dense in that commutant.

Due to example 3°/ in the introduction, we can now put forward

Proposition 10.2. Let A be ultraweakly closed with condition II and cofinal abelian sequence Δ_i, and $M = A_{id}$ be standard (i.e., J is an anti-unitary involution, such that $\Phi : T \in M \to JTJ \in M'$ is an anti-isomorphism). Let $\nabla_i = J\Delta_i J$ for $i \geq 0$ and $\mathcal{D}_{(\nabla)} = J\mathcal{D}_{(\Delta)}$.

1°/ $JAJ \equiv \cup_{n\geq 0} M'\nabla_n \times \nabla_n$ is ultraweakly closed, with condition II, relative to $\mathcal{D}_{(\nabla)} \hat{\otimes} \mathcal{D}_{(\nabla)}$.

2°/ The map Φ extends uniquely as an anti-isomorphism from A onto JAJ and JAJ is the commutant of A calculated in $B(\mathcal{D}_{(\nabla)}, \mathcal{D}_{(\nabla)})$.

This shows that a faithful normal linear form $f \geq 0$ on A, with G.N.S. representation $(\pi_f, H_f, \mathcal{D}_f, \zeta_f)$, induces a privilegiate choice of the commutant of $\pi_f(A)$. Indeed, we take the involution J_f of the left Hilbert algebra (A_{id}, f) and we point out [23] that J_f is the unique involution J associated to an anti-isomorphism from $\pi_f(M)$ onto $\pi_f(M)'$, with properties $J\zeta_f = \zeta_f$ and $JTJ = T^*$ for $T \in \pi(M) \cap \pi(M)'$. This commutant $J_f\pi_f(A)J_f$ for $\pi_f(A)$ is the ultraweak closure relative to $J(\mathcal{D}_f) \hat{\otimes} J(\mathcal{D}_f)$ of the right-regular representation of $A \cap L(\mathcal{D})$ in remark 7.1.

CHAPTER 11: ON STRONG AND ULTRASTRONG TOPOLOGIES

The terms 'σ-strong', 'strongest', and 'ultrastrong' have the same meaning, and are used alternatively in the following treatment.

Definition 11.1. Let A be a linear subspace of $B(\mathcal{D}, \mathcal{D})$, stable under involution, and $\mathfrak{S}$ be the family of bounded subsets of the Fréchet space $\mathcal{D}$. We define the strong topology on A to be the topology on A associated to the semi-norms:

$$T \in A \to p_{x_1,\dots,x_n;\, B_1,\dots,B_n}(T) = \sup_{y_1\in B_1,\dots,y_n\in B_n} \left| \sum_{i=1}^{n} (Tx_i, y_i) \right|$$

with $x_1,\dots,x_n \in \mathcal{D}$, $B_1,\dots,B_n \in \mathfrak{S}$, n moving in $\mathbb{N}$.

It is useful to give an alternative definition of such a topology. Let n be a fixed integer : for $B_1,\dots,B_n \in \mathfrak{S}$, the set $B = B_1 \times B_2 \times \cdots \times B_n$ may be identified to a bounded set of the Fréchet space $\mathcal{D}_n \equiv \mathcal{D}\oplus\cdots\oplus\mathcal{D}$ (n times), this last linear set being viewed as a dense subset of the Hilbert space $H_n \equiv H\oplus\cdots\oplus H$ (n times). The bounded sets $\pi_{j=1}^{n} B_j$, with $B_j \in \mathfrak{S}$, are a fundamental system of bounded subsets of $\mathcal{D}_n$, so that strong topology may be defined by the semi-norms

$$T \in A \to \left|\sup_{y\in B} ((T \otimes 1_{H_n})x,\, y)\right|,$$

with $x = (x_j)_{1\le j\le n} \in \mathcal{D}_n$, and B moving in the family of bounded subsets of $\mathcal{D}_n$, with $n \geq 0$.

Thus, when $\mathcal{D}$ is the Hilbert space H itself, taking for B_i the unit ball of the Hilbert space $H_i = H$ for $1 \le i \le n$, we get

$$p_{x_1,\dots,x_n;\, B_1,\dots,B_n}(T) = \sum_{i=1}^{n} \|Tx_i\|$$

and, taking for B the unit ball of $H_n = \bigoplus_{i \le j \le n} H_j$, we find that

$$\left(\sum_{i=1}^{n} \|Tx_i\|^2\right)^{1/2} = \sup_{\substack{y=(y_1,\dots,y_n) \\ \|y\| \le 1}} |(Tx_1, y_1) + \cdots + (Tx_n, y_n)|,$$

a well-known quantity. Thus,

Definition 11.2. Let $\mathcal{A}$ be a linear subspace of $B(\mathcal{D}, \mathcal{D})$, stable under involution, H_0 be a Hilbert space with countable Hilbert dimension, and $\Phi : T \in \mathcal{A} \mapsto T \otimes 1_{H_0} \in \mathcal{A} \otimes \mathbb{C}_{H_0}$ be the natural ampliation. The strongest (or σ-strong) topology on $\mathcal{A}$ is the topology on $\mathcal{A}$ which makes Φ bicontinuous when $\mathcal{A} \otimes \mathbb{C}_{H_0}$ is endowed with strong topology.

It has been seen that $\mathcal{A} \otimes \mathbb{C}_{H_0}$ is a space of continuous sesquilinear forms on the domain D of all σ-convergent sequences $x = (x_i)_{i \in \mathbb{N}}$, with values in $\mathcal{D}$, and D is easily seen to be a Fréchet space. If B is a bounded subset of D, one has, for every $\beta \in \mathcal{A}$ or $\beta \in B(\mathcal{D}, \mathcal{D})$,

$$\left|\sum_{i=1}^{+\infty} \beta(x_i, x_i)\right| \le m_\beta < +\infty$$

independently of $x = (x_i)_{i \ge 0}$ in B; therefore,

$$\sum_{i=1}^{+\infty} \|Ax_i\|^2 \le m_A < +\infty$$

for every operator $A \in \mathcal{A}$, or $A \in L(\mathcal{D})$ - it suffices to take for A any operator A_j in a given cofinal sequence in $\mathcal{A}^+$ - which is obviously equivalent to

$$\sup_{y \in B} |(A \otimes I_{H_0})(x, y)| \le m_A^2.$$

We point out that every bounded set B of the Fréchet space D induces on each $H_j = H$ a bounded subset $B_j = \pi_j(B)$ of the Fréchet space $\mathcal{D}$, where $\pi_j : \oplus_{K \ge 1}^{\infty} H_k \to H_j$ is the j-th projection, so that $B \subset \pi_{k=1}^{\infty} \pi_k(B)$ (this last set is not necessarily included in D).

From proposition 8.2°/ of [10], ρ-topology on A is finer than the strong and the σ-strong topologies. In fact, if B is a bounded subset of (A, strong topology) on (A, σ-strong topology) B(x, y) is uniformly bounded for each x and y in $\mathcal{D}$, implying that B must be bounded in (A, ρ), i.e., bounded sets in (A, ρ) (A, strong) and (A, σ-strong) coïncide.

Finally, the strong and strongest topologies coïncide on bounded sets of (A, ρ). Indeed, it suffices to take a σ-strong neighborhood W of zero in A of the form

$$W = \left\{\beta \in A \,\middle|\, \left|\sum_{\alpha=1}^{\infty} \beta(x_\alpha, y_\alpha)\right| \leq 1 \quad \text{for all } y = (y_\alpha) \in B\right\},$$

where $x = (x_\alpha)$ is a fixed vector in D, and B is a fixed bounded subset of D, noting that any σ-strong neighborhood of zero contains a finite intersection of neighborhoods such as W. As $(A_j^2 \otimes 1)(B)$ is bounded, we may assume that $\Sigma_{\alpha=1}^{\infty} \|A_j y_\alpha\|^2 \leq 1$ for all $y = (y_\alpha) \in B$. Let

$$V = \left\{\beta \in A \,\middle|\, \left|\sum_{\alpha=1}^{n} \beta(x_\alpha, y_\alpha)\right| \leq 1 \quad \text{for all } y = (y_\alpha) \in B\right\}$$

be the strong neighborhood of zero in A, where n is an integer chosen such that $\Sigma_{\alpha=n+1}^{\infty} \|A_j x_\alpha\|^2 \leq 1/16$. Taking a bounded set C in (A, ρ) of the form $C = [-A_j^2, A_j^2] \oplus i[-A_j^2, A_j^2]$, we see that any $\beta = \beta_1 + i\beta_2 \in V \cap C$ satisfies, for $y = (y_\alpha) \in B$,

$$\left|\sum_{\alpha=1}^{\infty} \beta(x_\alpha, y_\alpha)\right| \leq \left|\sum_{\alpha=1}^{n} \beta(x_\alpha, y_\alpha)\right| + \left|\sum_{\alpha=n+1}^{\infty} (BA_j x_\alpha, A_j y_\alpha)\right|,$$

since β can be written $\beta = BA_j \times A_j$, with

$$B = B_1 + iB_2 \in L(H) \qquad \|B_1\| \leq 1, \; \|B_2\| \leq 1.$$

The first term is $\leq 1/2$, and the second is smaller than $(\Sigma_{\alpha=n+1}^{\infty} \|B\| \; \|A_j x_\alpha\|^2)^{1/2}$, which itself is smaller than $2(\Sigma_{\alpha=n+1}^{\infty} \|A_j x_\alpha\|^2)^{1/2} \leq 1/2$, hence $V \cap C \subset W$, and our assertion follows.

Remark 11.1. 1°/ The reader can also check the following fact: let $\mathcal{D}_0$ be a dense linear subspace of the Fréchet space $\mathcal{D}$, and let $j \geq 0$ be fixed. On the bounded set $[-A_j, A_j]$ of

$(\mathcal{A}, \rho)$, the strong topology of $\mathcal{A}$ coïncides with the topology of semi-norms $q_{x,B} : \beta \in \mathcal{A} \mapsto q_{x,B}(\beta) = \sup\{|\beta(x, y)| \; y \in B\}$, where x moves in $\mathcal{D}_0$ and B in the family of bounded sets in $\mathcal{D}_0$.

2°/ Left and right multiplication

$$\beta \in B(\mathcal{D}, \mathcal{D}) \mapsto \beta(S \cdot , \cdot) \in B(\mathcal{D}, \mathcal{D})$$

and

$$\beta \in B(\mathcal{D}, \mathcal{D}) \mapsto \beta(\cdot , S \cdot) \in B(\mathcal{D}, \mathcal{D})$$

by an element $S \in L(\mathcal{D})$ are strongly and σ-strongly continuous. The multiplication $(S; T) \in L(\mathcal{D}) \times L(\mathcal{D}) \mapsto ST \in L(\mathcal{D})$ is strongly and σ-strongly continuous when S is restricted to move in a bounded set of $(L(\mathcal{D}), \lambda)$. Involution is not continuous for these topologies.

Proposition 11.1. Let $\mathcal{A}$ be a space with a cofinal central sequence. On every bounded set of $(\mathcal{A}, \rho)$, the strong - and ultrastrong - topology of $\mathcal{A}$ coïncides with the topology defined by the semi-norms $p_{x_1,\dots,x_n}$ with

$$p_{x_1, \dots, x_n} : T \in \mathcal{A} \mapsto p_{x_1, \dots, x_n}(T) = \left(\sum_{i=1}^{n} \|Tx_i\|^2\right)^{1/2}$$

where $(x_1, \dots, x_n)$ moves in the family $\mathfrak{S}_0$ of finite sets in the Fréchet space $\mathcal{D}$.

Proof. By the Hahn-Banach theorem, we are reduced to the case where $\mathcal{A} = \cup_{i\geq 0} \mathcal{A}_{A_i}$ satisfies condition II, and A_j being central, we may assume that $\mathcal{A}$ is ultraweakly closed. For a bounded set C of $(\mathcal{A}, \rho)$, being contained in an order-interval, we may assume that $C = [-A_j, A_j]$ for some j and, by proposition 11.2, we are reduced to the case where $\mathcal{A} = \mathcal{A}_j$. Let H_1 be the unit ball of H, $x \in \mathcal{D}$ and $\bar{A}_j = \int_1^\infty \lambda \, dp_\lambda$ the spectral theorem for $\bar{A}_j$. Putting $E_n = \int_1^n dp_\lambda$, one has, for $T \in C$ and $T \to 0$, for strong topology,

$$\begin{aligned} |(Tx, y)| &\leq |(Tx, E_n y)| + |(Tx, (1 - E_n)y)| \\ &\leq |(Tx, E_n y)| + \|Tx\| \, \|(1 - E_n)y\|. \end{aligned}$$

The first term tends to zero by lemma 3.1. As the set of T^2, $T \in C$, remains bounded for $\lambda = \rho$ topology (corollary 6.1), we find some $k \geq 0$, such that $\|Tx\| \leq C^{te}\|A_j^k x\|$ for all $x \in \mathcal{D}$, thus the second term tends to zero as $E_n \to \mathrm{Id}$ and the proposition follows from $\|Tx\| = \sup\{|(Tx, y)| \; ; \; y \in H_1\}$.

In practice, strong topology or σ-strong topology depend on the choice of a fundamental system of bounded subsets in the Fréchet space $\mathcal{D}$, and it is a known fact that the metrisability of $\mathcal{D}$ implies that a fundamental sequence of bounded subsets in $\mathcal{D}$ (when $\mathcal{D} \neq H$) never exists. Let us consider a space A of the form $A = \cup_{n\geq 0} A_{\Delta^n}$, $\Delta \geq 1$, with the spectral representation $\bar{\Delta} = \int_1^\infty \lambda \, dp_\lambda$, and let us denote (for $n \geq 1$, $n \in \mathbb{N}$) by E_n the sequence of projectors $E_n = \int_1^n dp_\lambda$, which clearly tends to identity for ρ-topology of A (as $n \to \infty$). By lemma 3.1, the unit ball B_n of each Hilbert space $E_n(H)$ is, for every $n \geq 0$, a bounded subset of the Fréchet space $\mathcal{D}$, and:

Lemma 11.1. The strong topology of $A = \cup_{n\geq 0} A_{\Delta^n}$ (with $\Delta \geq \mathrm{Id}$) coïncides on bounded subsets of A with the topology generated by the semi-norms

$$q_{x, B_n} : \beta \in A \to q_{x, B_n}(\beta) = \sup\{|\beta(x, y)|; \quad y \in B_n\},$$

with x moving in $\mathcal{D}$ and n in $\mathbb{N}$.

Proof. Let $\ell \geq 0$ fixed, and $[-\Delta^\ell, \Delta^\ell]$ be the corresponding bounded set of (A, ρ) under consideration. It suffices to show that a strong neighborhood W of the form

$$W = \{\beta \in A | \sup_{y \in B} |\beta(x, y)| \leq 1\},$$

where x is in $\mathcal{D}$ and B is a bounded set in $\mathcal{D}$, induces on $[-\Delta^\ell, \Delta^\ell]$ a zero neighborhood associated to a suitable q_{x, B_n}. For $\beta \in A$, we put $\beta_\alpha = \beta E_\alpha \times E_\alpha$. There exists an integer k (one may choose $k = 4\ell + 2$) and a sequence $\varepsilon(\alpha) \to 0$ (as $\alpha \to \infty$) such that, for all $u \in \mathcal{D}$,

$$|(\beta - \beta_\alpha)(u, u)| \leq \varepsilon(\alpha)(\Delta^k u, u)$$

whenever $|\beta| \leq \Delta^\ell$. From polarization equality, we get

$$|(\beta - \beta_\alpha u, v)| \leq \varepsilon(\alpha)((\Delta^k u, u) + (\Delta^k v, v))$$

for $u, v \in \mathcal{D}$, so that we may choose $n \in \mathbb{N}$, such that

$\varepsilon(n)[(\Delta^k B, B) + (\Delta^k x, x)] \leq 1/2$ since $(\Delta^k B, B)$ is bounded in $\mathbb{R}$. We may assume that $E_n(B) \subset B_n$, and associate to the semi-norm qx, B_n the strong neighborhood

$$V = \{\beta \in A \mid \sup_{y \in B_n} \mid \beta(E_n x, y) \mid \leq 1/2\}.$$

Then, for $\beta \in [-\Delta^\ell, \Delta^\ell] \cap V$, one has, with $y \in B$,

$$|\beta(x, y)| \leq |(\beta - \beta_n)(x, y)| + |\beta_n(x, y)|$$

$$\leq 1/2 + |\beta(E_n x, E_n y)| \leq 1,$$

i.e., $W \supset [-\Delta^\ell, \Delta^\ell] \cap V$, thus implying the lemma.

Proposition 11.2. Let $A = \cup_{i \geq 0} A_{A_i}$ and $j \geq 0$ fixed, $A_j = \cup_{n \geq 0} A_{id} A_j^n \times A_j^n$ with natural domain $\mathcal{D}_j = \cap_{n \geq 0} \bar{A}_j^{-n}(H)$. On the bounded set $[-A_j, A_j]$ of (A, ρ), the strong topology of A coïncides with the strong topology of A_j (which refers to the Fréchet space $\mathcal{D}_j$). If the Hilbert space H is separable, the strong topology of A restricted to a bounded set in (A, ρ) is metrizable.

Proof. By the Hahn-Banach theorem, we are reduced to $A = B(\mathcal{D}, \mathcal{D})$. Replacing Δ by A_j in the proof of lemma 11.1, and noting that $\mathcal{D}$ is dense in the Fréchet space $\mathcal{D}_j$ (by proposition 3.5), it suffices to apply the remark associated to definition 11.2. By proposition 3.7, the separability of H is equivalent to the separability of the Fréchet space $\mathcal{D}_j$ (or of $\mathcal{D}$), so that our assertion of metrisability rests on lemma 11.1 and the remark just made.

Theorem 11.1. Let A be a subspace of $B(\mathcal{D}, \mathcal{D})$, stable under involution. The dual of A endowed with strong topology (respectively σ-strong topology) consists of weakly (respectively σ-weakly) continuous - relative to $\mathcal{D} \hat{\otimes} \mathcal{D}$ - linear forms on A.

Thus, for an ultraweakly closed A, the dual of (A, σ-strong topology) endowed with its natural strong topology (i.e., topology of uniform convergence on σ-strongly bounded subsets of A) is exactly the Fréchet space P_A, the predual of A.

Proof. Let f be a strongly (respectively σ-strongly) continuous linear form on A. By the Hahn-Banach theorem, we are reduced to $A = B(\mathcal{D}, \mathcal{D})$ and, introducing a suitable ampliation of A, we are reduced to the case where

$$|f(T)| \leq M \sup_{y \in B} |T(x, y)|$$

for all $T \in B(\mathcal{D}, \mathcal{D})$, with x a fixed vector in $\mathcal{D}$ and B a bounded set in $\mathcal{D}$ ($M < +\infty$ will be chosen $\equiv 1$). The restriction of f to each $B(\mathcal{D}, \mathcal{D})_{A_i^2}$ with the usual notation $B(\mathcal{D}, \mathcal{D}) = \cup_{i\geq 0} B(\mathcal{D}, \mathcal{D})_{A_i^2}$, is Hilbert strongly continuous on $L(H)$, since one has, for $\beta = BA_i \times A_i$ with $B \in L(H)$,

$$|f(BA_i \times A_i)| \leq \sup_{y \in B} |(BA_i x, A_i y)| \leq C^{te} \|A_i x\|,$$

noting that $A_i(B)$ is a bounded set in the Hilbert space H, as A_i induces a continuous linear map from $\mathcal{D}$ Fréchet into H Hilbert. Thus, this yields a vector $y_i \in H$ such that, for all $B \in L(H)$,

$$f(BA_i \times A_i) = (BA_i x, y_i)$$

and, from $\bar{A}_i(\mathrm{Dom}\, \bar{A}_i) = H$, we find that $\zeta_i \in \mathrm{Dom}\, (\bar{A}_i)$, such that $y_i = \bar{A}_i \zeta_i$. It remains to show that ζ_i is independent of i or, equivalently, that $\zeta_i = \zeta_0$, where ζ_0 satisfies, for all $B \in L(H)$,

$$f(B) = (Bx, \zeta_0) = (BA_0 x, \bar{A}_0 \zeta_0),$$

due to $A_0 = \mathrm{Id}$. Let i be fixed and $\mathcal{D}_i = \cap_{n\geq 0} \bar{A}_i^{-n}(H)$ be the natural domain of $\cup_{n\geq 0} L(H) A_i^n \times A_i^n$. Denote by F_i the space of finite dimensional rank operators from H into $\mathcal{D}_i$. For $u \in \mathcal{D}_i$, $v \in \mathcal{D}_i$, the operator $B(\cdot) = \langle \cdot\, , u\rangle v$ is in F_i and the form $BA_i \times A_i$ corresponds to the operator $BA_i \times A_i = \langle \cdot\, , A_i u\rangle A_i v$, implying that $F_i A_i \times A_i \subset F_i$ (since $\bar{A}_i(\mathcal{D}_i) = \mathcal{D}_i$). When u and v are in H, the operator $B_\alpha(\cdot) = \langle \cdot\, , E_\alpha u\rangle E_\alpha v$ is in F_i where E_α, for $\alpha \geq 1$, is the projector $E_\alpha = \int_1^\alpha dp_\lambda$ associated to the spectral representation $\bar{A}_i = \int_1^\infty \lambda\, dp_\lambda$ of the self-adjoint operator $\bar{A}_i$. From $B = B_\alpha A_i \times A_i \in F_i \subset L(H)$, we obtain

$$f(B) = f(B_\alpha A_i \times A_i) = \langle x, A_i E_\alpha u\rangle\langle A_i E_\alpha v, \zeta_0\rangle$$

and, on the other hand,

$$f(B_\alpha A_i \times A_i) = \langle A_i x, E_\alpha u\rangle\langle E_\alpha v, A_i \zeta_i\rangle,$$

leading to $\zeta_i = \zeta_0$, since u and v are arbitrary chosen in H. It follows that $f(\beta) = (\beta x, \zeta_0)$, i.e., the weak continuity of f.

To sum up the proposition, let $\varphi : \mathcal{A} \to \mathbb{C}$ be a linear form. Then φ is weakly continuous $\Leftrightarrow$ φ is strongly continuous $\Leftrightarrow$ $\varphi = \Sigma_{i\in I}\, \omega_{x_i,y_i}$ with $x_i \in \mathcal{D}$, $y_i \in \mathcal{D}$ and I finite. In the same manner, φ is σ-weakly continuous $\Leftrightarrow$ φ is σ-strongly continuous $\Leftrightarrow$ $\varphi = \Sigma_{i=1}^{\infty}\, \omega_{x_i,y_i}$, with $x = (x_i)_{i\in\mathbb{N}}$ and $y = (y_i)_{i\in\mathbb{N}}$ σ-convergent sequences in $\mathcal{D}$.
We immediately get:

Corollary 11.1. For a subspace $\mathcal{A} = \cup_{i\geq 0}\, \mathcal{A}_{A_i}$ of $B(\mathcal{D}, \mathcal{D})$, stable under involution, its weak closure (respectively σ-weak closure) is identical to its strong closure (respectively σ-strong closure).

Finally:

Proposition 11.3. Let $\mathcal{A} = \cup_{i\geq 0}\, \mathcal{A}_{A_i}$ be a subspace of $B(\mathcal{D}, \mathcal{D})$, stable under involution and ultraweakly closed. Then:

1°/ A linear form $\varphi : \mathcal{A} \to \mathbb{C}$ is σ-strongly continuous iff its restriction to each $[-A_i, A_i]$ is strongly continuous for every $i \geq 0$.

2°/ For a convex set K in $\mathcal{A}$, the following conditions are equivalent:

2.1. K is σ-weakly closed;

2.2. K is σ-strongly closed;

2.3. $K \cap [-A_i, A_i]$ is σ-weakly (or weakly) closed, for every $i \geq 0$;

2.4. $K \cap [-A_i, A_i]$ is σ-strongly (or strongly) closed, for every $i \geq 0$.

Proof. Let $P_{\mathcal{A}}$ be the Fréchet space predual of $\mathcal{A}$. Bounded sets of $\mathcal{A}$ with σ-strong topology are identical to bounded sets of $(\mathcal{A}, \rho)$, so that the strong dual P of $(\mathcal{A}, \sigma\text{-strong})$ is a topological linear subspace of the Fréchet space $\mathcal{A}^\rho$, and, due to theorem 11.1, $P = P_{\mathcal{A}}$ topologically. By a theorem attributed to Grothendieck, the completion of P is the space of linear forms $\varphi : A \to \mathbb{C}$, whose restrictions to each bounded sets of A is σ-strongly continuous, or, equivalently, strongly continuous (due to their coïncidence on the intervals $[-A_i, A_i]$), so that first assertion follows from completeness of P_A^i.

By Minkowski's theorem, the notion of a closed convex set K, for a given locally convex topology C on A, depends only on the dual system $\langle A, A' \rangle$ where A' is the dual of (A, C) : this yields 2.1 ⇔ 2.2 and 2.3 ⇔ 2.4. The equivalence 2.1 ⇔ 2.3 is the Banach-Dieudonné theorem.

Proposition 11.4. Let A and B be ultraweakly closed spaces satisfying condition II, and $\pi : A \to B$ a normal homomorphism from A onto B. Then the restriction of π to bounded sets of (A, ρ) is strongly continuous. If π is an isomorphism, π is σ-strongly bicontinuous from A onto B.

Proof. Since π is the normal lifting of its normal restriction to A_{id}, we are reduced to the case where π is either an ampliation or a reduction, the case π unitary implemented being obvious. Let $T \in A \to T \otimes Id \in A \otimes \mathbb{C}_K$ be an ampliation, where K is a given Hilbert space. By definition 11.1, Φ is bicontinuous from $(A, \sigma\text{-strong})$ onto $A \otimes \mathbb{C}_K$ strong, and our assertion is the coïncidence of σ-strong and strong topologies on bounded sets in (A, ρ). Now, let $\Phi : \beta \in A = \cup_i A_{A_i} \to \beta E{\times}E \in A_E$ be a reduction, where E is a given projector in A'. For $\beta \in [-A_j, A_j]_A$, i.e., $\beta \in [-A_j, A_j]_{A_j}$ strongly tending to zero in A (i.e., $\beta(x, B) \to 0$ for all $x \in \mathcal{D}_j$ and B bounded set in $\mathcal{D}_j$), we find that $\beta(Ex, E(B)\ \mathcal{D}{\to}0$, since $Ex \in \mathcal{D}$ and $E(B)$ is bounded in $\mathcal{D}$, hence our assertion follows from proposition 4.1, for example. Now, let π be an isomorphism and $g : B \to \mathbb{C}$ be a σ-strongly continuous linear form. Then $h = g \circ \pi : A \to \mathbb{C}$ is a linear form whose restriction to bounded sets of (A, ρ) is strongly continuous. By proposition 11.3, h is σ-strongly continuous, thus proving the proposition.

Lemma 11.2. For a space A with condition II, its weak closure, its strong closure, its σ-weak closure, and its σ-strong closure in $B(\mathcal{D}, \mathcal{D})$ are all identical.

Proof. By Minkowski's theorem and proposition 11.3, the weak closure (respectively σ-weak closure) coïncides with the strong (respectively σ-strong) closure. It is now clear that $A \subset A^\sigma \subset A^w \subset B(\mathcal{D}, \mathcal{D})$ where w stands for weak topology, thus we need to show that A^σ is weakly closed. As A^σ satisfies condition II, its bounded part $(A^{\sigma})_{id}$ is a von Neumann algebra = M with property $M'\mathcal{D} \subset \mathcal{D}$. The set A_1 of $\beta \in B(\mathcal{D}, \mathcal{D})$, satisfying

$$\beta(Bx, y) = \beta(x, B^*y)$$

for all $B \in M'$ and all $x, y \in \mathcal{D}$, is clearly weakly closed in $B(\mathcal{D}, \mathcal{D})$ and theorem 1.2 states that $A_1 = A^{\sigma}$, hence the lemma.

Theorem 11.2 (Kaplansky's density theorem. Let $\mathcal{A}$ and $\mathcal{B}$ be spaces with condition II and same cofinal sequence A_i. Let $\mathcal{A}$ be strongly dense in $\mathcal{B}$, with $\mathcal{A} \subset \mathcal{B}$. Then, for every integer i, the interval $[-A_i, A_i]_{\mathcal{B}}$ is strongly dense in the interval $[-A_i, A_i]_{\mathcal{B}}$.

REFERENCES

[1] H. Araki and J.P. Jurzak: 'On a certain class of ⋆-algebras'. Publ. RIMS, Kyoto Univ. 18 (1982), 1013-1044.

[2] H. Borchers: 'Energy and momentum as observables in quantum field theory'. Comm. Math. Phys. 2 (1966), 49-54.

[3] J. Dieudonne and L. Schwartz: 'La dualité dans les espaces (F) et (LF)'. Ann. Inst. Fourier (Grenoble) 1 (1949), 61-108.

[4] J. Dixmier: Les algèbres dópérateurs dans l'espace hilbertien. 2nd Ed. Gauthier-Villars, Paris, 1969.

[5] A. Connes: 'A survey of foliations and operator algebras'. (preprint 1981, IHES).

[6] N. Dunford and J.T. Schwartz: Linear operators, Part II. Interscience Publishers, 1963.

[7] A. Grothendieck: Topological vector spaces. Gordon and Breach, New York, 1973.

[8] A. Grothendieck: 'Produits tensoriels topologiques et espaces nucléaires'. Mém. Amer. Math. Soc. 16 (1955).

[9] G. Jameson: Ordered linear spaces. Springer-Verlag, Berlin 1970.

[10] J.P. Jurzak: 'Unbounded operator algebras and DF-spaces'. Publ. RIMS Kyoto Univ. 17 (1980), 755-776.

[11] J.P. Jurzak: 'Propriétés topologiques des algèbres d'opérateurs non bornés'. Thesis.

[12] T. Kato: Perturbation theory for linear operators. Springer, Berlin, 1966.

[13] I. Kovacz and J. Szücs: 'Ergodic type theorems in Von-Neumann algebras'. Acta. Sc. Math. 27 (1966), 233.

[14] J. Moser: 'A new technique for the construction of solutions of nonlinear differential equations'. Proc. N.A.S., 1824-1830 (1962).

[15] E. Nelson: 'Analytic vectors'. Ann. Math. 70 (1959), 572.

[16] E. Nelson and W.F. Stinespring: 'Representations of elliptic operators in an enveloping algebra'. Amer. J. Math. 81 (1959) 547-560.

[17] J.V. Neumann and F.J. Murray: 'On rings of operators'. Ann. Math. 37 (1936) 116-229.

[18] J. Poulsen: 'On C^∞-vectors and intertwining bilinear forms for representations of Lie groups'. J. Funct. Analysis 9 (1970) 87-120.

[19] R.T. Powers: 'Self-adjoint algebras of unbounded operators'. Comm. Math. Phys. 21 (1971) 85.

[20] S. Sakaï: C^*-algebras and W^*-algebras. Springer, Berlin, 1971.

[21] I.E. Segal: 'A non-commutative extension of abstract integration'. Ann. Math. 57 (1953) 401-457.

[22] T. Sherman: 'Positive functionals on $*$-algebras of unbounded operators'. J. Math. Anal. Appl. 22 (1968) 285-318.

[23] S.L. Woronowicz: 'On the purification of factor states'. Comm. Math. Phys. 28 (1972).

[24] S.L. Woronowicz: 'The quantum problem of moments - II'. Reports Math. Phys. 1 (1971) 175-183.

SUBJECT INDEX